Outdoor Survival

Garth Hattingh

First published in 2003 by
New Holland Publishers Ltd
London • Cape Town • Sydney • Auckland
www.newhollandpublishers.com

86 Edgware Road
London, W2 2EA
United Kingdom

80 McKenzie Street
Cape Town, 8001
South Africa

14 Aquatic Drive
Frenchs Forest, NSW 2086
Australia

218 Lake Road
Northcote, Auckland
New Zealand

ISBN 1 84330 254 3 (hardback)
ISBN 1 84330 255 1 (paperback)

Publisher and Editor: Mariëlle Renssen
Publishing Manager: Claudia Dos Santos
Managing Editor: Simon Pooley
Managing Art Editor: Richard MacArthur
Designer: Sheryl Buckley
Illustrator: Steven Felmore
Picture Researcher: Bronwyn Allies
Production: Myrna Collins
Consultant: Dr Lance Michell

Reproduction by Unifoto Pty Ltd
Printed and bound in Singapore by
Craft Print International Ltd

2 4 6 8 10 9 7 5 3 1

Disclaimer

The author and publishers have made every effort to ensure that the information contained in this book was accurate at the time of going to press, and accept no responsibility for any injury or inconvenience sustained by any person using this book or following the advice herein.

Author's acknowledgments

My most heartfelt thanks go to Dr Michell, who brought his extensive experience in mountain rescue, survival medicine as well as trauma management to bear in the medical chapter of this book. In addition, he also acted as general consultant on many other sections, ensuring that the latest information and techniques are outlined. Thanks are also due to the many real survivors who have put their experiences into print to encourage and give hope to others who may, some day, find themselves in dire trouble. Search for survival accounts in your library and be touched by them, like I was. To Willem, Tessa, Johan, John and Frans – thank you for your patient support, as well as your help with photo shoots. Last, but far from least, to the team at New Holland Publishers in Cape Town, in particular Mariëlle Renssen, thank you all for your friendly help.

Outdoor Survival

Contents

The Art of Survival

How does it happen, in our high-tech world of 'comfort zones', that on many occasions people find themselves in a desperate situation in the remote wilderness?

These predicaments are often unavoidable – via an aeroplane, bus or car crash, a derailment or a shipwreck. They could stem from unexpected natural phenomena, such as an earthquake, hurricane or exceptionally violent storm.

However, the vast majority of survival situations are a result of human fallibility – largely due to the tendency of people to enter into outdoor activities without thorough preparation. The growing popularity in hiking, adventure travel and activities such as camping, canoeing, mountain biking, and even rock and ice climbing has seen a huge increase in the number of incidents in wilderness areas. Surprisingly few of these involve the macho, wild-man extremists who push the limits of the possible, and whom one would expect to land in trouble. Rather, most serious survival situations arise out of ordinary, unassuming family drives, boating trips, walks or treks into backwood areas, where the much-maligned Boy Scout motto, Be Prepared, is ignored. They occur because of insufficient forethought, lack of essential equipment, or simple inexperience and insufficient knowledge of outdoor principles. The same people who meticulously plan a shopping trip casually embark on wilderness adventures with scarcely any pre-planning.

The answer to all of this is: Get prepared!

opposite ICE CLIMBING IS ONE ACTIVITY WHERE PREPARATION AND TRAINING ARE PARAMOUNT, AS ONE MISPLACED FOOTING COULD SPELL DISASTER. BESIDES THE METICULOUS PLANNING, CORRECT EQUIPMENT, SUCH AS CRAMPONS, PICK AXE AND HELMET, IS INDISPENSABLE.

Survival essentials

Never underestimate the wilderness. Many groups venture forth with no knowledge of the terrain they are heading into and no prior training in the adventure activities they intend doing, such as canoeing, climbing or river rafting. Good planning – which includes research on the area, suitable equipment, activity experience and adequate physical fitness – will help make the outdoor adventure safe and enjoyable.

Part of being prepared is learning about and practising outdoor and survival skills such as map reading, navigation, creating shelters, river crossing, finding food and water, making fires and first-aid principles. Many of the techniques (such as trapping) should only be applied in genuine survival situations, but 'mock runs' are invaluable in the event of a crisis.

Avoiding survival situations

The obvious thing is to avoid getting into a survival situation but there are times when you can do nothing to prevent it – such as unpredictable natural disasters. However, where you can, the trick is to be adequately prepared for any outdoor excursion or activity, and to be sensible about what you attempt. Match the level of the activity to the experience and equipment of the group. You need to ask yourself two key questions.

One: do you have sufficient background and knowledge for the task? There is no substitute for experience; going deep into back-country in the middle of winter is hardly a suitable activity for novice hikers, yet an amazing number of rookie groups do just this, often with severe consequences.

Two: do you have the right equipment for the trip? Ill-equipped groups often get into grave situations as a result of backpack and footwear problems or inadequate tents and clothing to survive the usual vagaries of the weather.

General preparedness

Always bank on a 'worst case' scenario when taking a trip into the great outdoors. Cater for the nastiest anticipated weather, and for the longest expected time, then add extra to both.

By reading a book such as this, you have gone part of the way; by practising some of the skills and improving your fitness, you are even better organized; by refining your equipment, you are prepared for all eventualities of weather and terrain; by researching and studying further, you are placing yourself in a position of strength and confidence should you ever need to apply the knowledge.

Planning your trip

Information is a vital key – the more facts you have on the area, the better prepared you are. Find out about weather patterns; the land form of mountains; direction and strength of river flow; vegetation; edible and inedible plants; animal life; ocean currents and water temperatures. The Internet and books offer a wealth of information, as do people who have already visited your particular area. Take up-to-date maps and guidebooks. Carefully plan the trip with your group.

Things of course seldom go totally according to plan. Depending on the nature of the trip, time and thought needs to go into the following.

ACTIVITIES SUCH AS ABSEILING, WHERE ONE MAY HAVE TO TRUST A SINGLE ROPE, EPITOMIZES THE MENTAL PREPAREDNESS AND FOCUSED CONCENTRATION NECESSARY FOR CERTAIN ADVENTURE ACTIVITIES.

Vital information

- **Draw up a schedule:** For all trips, especially those in back-country, the wilderness and at sea, have an expected departure and trip completion time, plus a list of contacts. Leave your schedule, dates, list of participants and contact details with someone trustworthy in case of an emergency.
  If you are not able to keep to your plan, the entrusted person(s) should be given clear instructions on what steps to take.
- **Route plan:** Leave a plan of your projected route with your contact person or with the relevant authorities such as forestry, mountain rescue, the harbour master, coastguard, or aviation control if you are undertaking a private flight.
- **Emergency escape routes:** If you are planning an extended trip in mountains or on a river, plot several emergency escape routes on a route card. Leave a copy with your contact person. This will help rescuers narrow their search options.
- **Regrouping plan:** For travel in back-country areas, make clear plans both beforehand and on a daily or even hourly basis for regrouping, or some other arrangement in case your group becomes separated or one of the members goes missing.

Leadership

Firm but tactful leadership can be essential in tough survival situations, particularly in the initial stages after a disaster. Depression, hopelessness, recklessness and confusion can all take their toll, mentally and physically. If a naturally competent leader does not emerge, then elect a 'captain' as soon as possible.

The ideal leader

- Shows confidence and optimism, even though he or she may not truly feel it.
- Is sensitive to group and individual needs, but able to exercise firm control over factors such as supply rationing; allocation of space, clothing and equipment; medical priorities; and distribution of tasks.
- Makes all major decisions in consultation with the group to avoid anyone feeling left out or abandoned.
- Is flexible and adaptable, but does not show uncertainty when the final decision has to be made.
- Is prepared to hand over leadership under certain circumstances, for example, if someone has more medical knowledge.

Mental preparation

Your mind is your greatest survival tool. Those who have survived against the odds have not necessarily been the toughest, physically, or had the ideal equipment. The essence of true survivors has been mental readiness to handle unexpected challenges.

Many recorded feats of incredible endurance have been achieved – some examples include undergoing up to 75 days without any food, a week without water, surviving for days in freezing conditions, epic perseverance in blazing sun. In many cases, though, some members of the group have simply given up and let death overtake them, but the will to survive has kept others going. A common factor in such individuals has been unshakable optimism, often associated with a strong faith or with a specific purpose for living – a partner, children or parents. By focusing on these factors rather than the doom and gloom of the situation, they succeeded in enhancing their own and others' chances of survival.

Truly, the difference between life and death often lies with the mind. You need to fight the doubts, fight the tendency to say 'I will never make it'. Focus instead on the survival records of others. If they did it, you can too!

A WISE LEADER INVOLVES THE ENTIRE GROUP IN PLANNING, AS THIS WILL STAND THEM IN GOOD STEAD SHOULD A SERIOUS SITUATION ARISE IN THE WILDERNESS. IF EVERYONE FEELS INVOLVED IN CRUCIAL DISCUSSIONS, IT MAKES THE DIFFICULT DECISIONS EASIER TO ACCEPT.

Physical preparation

Although perhaps not the most important factor in survival, challenging any rugged outdoor environment without being fit is a form of idiocy. All too often, people end up risking their lives rescuing someone who has got into difficulties due to being unfit. Also ensure that health basics are taken care of before a long trip – dental problems, any nagging ailments, sprained muscles or ligament injuries. Have all required inoculations and purchase anti-malarial drugs or similar, if necessary, before your trip. The rest of physical preparedness involves appropriate exercise – aerobics, weight training, running, cycling, rowing. Use a consultant to work out a suitable training programme.

Many activities need very definite skills. Sea kayaking needs expertise in handling waves and currents (experiencing choppy conditions is not the time to learn how to eskimo roll!); preparation for negotiating sand desert might include 4x4 driving courses and motor maintenance. If you're on a climbing expedition, fumbling with unfamiliar knots on a mountain in a storm is not advised! Don't hesitate to consult experts for coaching and advice before your trip. Before going into the wilderness in a vehicle, check that it is mechanically sound, has a spare tyre, that you have the necessary tools to effect minor repairs, that the jack works, that you have snow chains and a shovel (in cold conditions), and that there is spare fuel, food and water. For a long drive in very cold country, ensure that there is warm clothing in the car – air conditioners and heaters are not much use when you are trapped in a blizzard. For desert driving, make sure you have plenty of spare water.

Work on the assumption that you might just have to overnight in the car, and pack accordingly.

For boats, carry out a similar procedure, including lifebelts or life rafts, flares, radio, emergency supplies, and general seaworthiness.

GOOD PHYSICAL FITNESS INCREASES YOUR CHANCES OF SURVIVAL WHEN VENTURING ON AN EXCURSION INTO THE WILDERNESS.

Choosing an adventure operator

Many have flawless records and safe operating procedures, others are less experienced and take chances, putting the group at risk. There are numerous examples of deaths and disasters resulting largely from operator error. It is worthwhile doing your homework thoroughly before you consider using the services of an adventure operator. Ask for recommendations from others who have done the same activity.

Your survival skills

Skills such as creating shelter, finding food, trapping, and moving through difficult and unfamiliar territory all require practice. In a survival scenario, you want to be familiar with the skills rather than be forced to use them for the first time.

Plan a family, club or scout troop survival weekend, complete with fire-making, building a shelter, some navigation exercises, and primitive fishing and food gathering. This can liven up a hiking trip or provide an entertaining time in the security of your local woods.

Be constantly aware of your position and surroundings. Do not rely only on the 'group leader' or 'guide' to follow your progress; ask to be shown where you are every time the map is consulted. Watch the lie of the land and know in which direction you are travelling. Make a mental note of features like rivers, lakes, roads and buildings; and of caves and hollows as possible shelter. This alertness will stand you in good stead if the need arises.

Key survival tip

STOP (**S** = Sit; **T** = Think; **O** = Observe; **P** = Plan)

Pausing to evaluate the resources of the situation and the group, and examining your options are essential before acting. Every survival situation is unique – a novel mix of people, a specific set of conditions, a differing amount of equipment. All this should be carefully weighed before deciding on a course of action.

Political hostility

Many areas of the world are subject to sudden political and/or religious turmoil, or even outright warfare, with little prior warning for the adventure traveller.

In being faced with hostile behaviour, be aware that the hostile parties will usually be armed and 'trigger-happy'. Take extreme care in how you react; avoid sudden hand movements, particularly into a bag, your vehicle or pockets, which could look like an attempt to reach for a weapon.

Avoid aggressive eye contact and any form of verbal argument or confrontation. Never shout and gesticulate; stay calm and friendly. Curb your impatience and annoyance – remember who holds the gun! Remain quiet and respectful, even if the demands are unreasonable. Make a submissive gesture, such as hands forward, palms out, head bowed. Carry a 'sacrificial bank roll' in a pocket or pouch where it can be easily reached. Your main cash and travel documents should be carried hidden in a thin pouch under your clothing.

Do not wear expensive clothes or jewellery and do not display money, expensive cameras and watches. These often represent more wealth than the average citizen of a poor country could possess in a lifetime. If someone is determined to take your possessions, hand them over. Your life should always take precedence.

Waiting may constitute part of the process. Petty border-post officials or military members like to delay, and may confiscate your goods or even imprison you if you show impatience.

Do not take photographs of military installations, government buildings or security forces.

Ensure that all legal medicines are properly boxed, with full labelling. NEVER carry drugs, and avoid carrying liquor, especially in Muslim countries. Respect local dress and cultural or religious codes of conduct (e.g. covering the head, women refraining from wearing shorts, not drinking in public).

IT IS CRUCIAL TO MAINTAIN A SENSE OF CALM WHEN FACED WITH HOSTILE BEHAVIOUR IN POLITICALLY TENSE CIRCUMSTANCES.

Get the Gear

Having the correct equipment, particularly certain basic survival items, could reduce or even avert the impact of a survival situation. How do you choose the best equipment from the vast array that is available? Research is the key. With each piece of equipment, convince yourself that some day you might really have to survive under extreme conditions. This will force you to spend time obtaining more detailed information. This should be done thoroughly. Sources include:

- Senior salespeople (but beware of profit motives)
- Internet
- Outdoor books and magazines
- Friends with similar gear.

Make sure you know exactly what you intend to use the items for (it is pointless buying a down sleeping bag designed to keep you comfortable at -30°C (-22°F) for a summer trip in the local hills). Make lists of suitable products, prices, performance statistics and outdoor-gear reviews from magazines. Armed with these, you can then buy more sensibly.

In general, the equation 'price = quality' does hold, although, fortunately, not always.

Whether you really have to survive with the kit or not, the exercise is still worthwhile.

Equipment for children

Many parents end up buying inferior equipment for children with the excuse: 'They soon grow out of it'. Keeping up with growth patterns can admittedly be an economic nightmare. On the other hand, one serious case of hypothermia (see pp80–81), or boots failing or a pack breaking and thus delaying a party, can give rise to a desperate survival scenario with exorbitant costs (rescue, hospitalization – or the saddest of all, personal trauma). As with adults, so with children: the wisest move is to buy the best you can. You can always sell or trade it off later with other families.

High-quality gear is often more lightweight than less expensive stuff, and children are not able to carry as much weight on a hike as adults. If anyone needs the new hi-tech stuff, it might just be the younger members of the family.

opposite CAREFUL FORWARD PLANNING NEEDS TO TAKE INTO ACCOUNT THE CORRECT TYPE OF CLOTHING TO SUIT YOUR CHOSEN ACTIVITY – NEVER UNDERESTIMATE ENVIRONMENTAL CONDITIONS.

Kids' survival pouch

Pack a kids' survival pouch using a small waist or belt pouch, and make the child wear it at all times when out in the wilderness. Teach your child how, when, and why to use each item.

Include in it:

Water bottle: Use only when really thirsty; best if attached to pouch.

Whistle: Plastic; blow ONLY when in trouble.

Space blanket: For shelter; be careful when unwrapping as it tears easily.

Lightstick or two: 12-hour variety; for security and visibility at night.

Bright scarf: Red or orange; as a flag, head-cover, or bandage.

Signal mirror: Metal or tough plastic; for signalling to a rescue party or aircraft.

Snacks: Sweets, chocolate; emergency only!

Mini first aid kit: plasters, bandage.

Actions card: One side has a reminder of the above points, the other the child's and your names, plus some contact numbers.

Family photo: Finally, including a picture of the family can offer great psychological comfort.

Survival skills for kids

- It is great to take youngsters into the outdoors, but make sure they, too, are prepared. Young children can easily get separated from their parents. Equip them with essential survival principles by discussing the possibility of their being lost and how to react to this, and by creating a special Kids' Survival Kit, which you discuss with your youngsters.
- **Stay together:** if you get lost with a friend or two, do not split up!
- **Stop moving:** as soon as you find you are lost or alone, STOP and sit under a tree or find a 'house' or 'fort' (NOT a hiding place but a sheltered space where your parents or searchers will still find you).
- **Mark your house:** use bright colours (such as the scarf from your survival kit) tied to a branch or on/under a rock.
- **Cover yourself** with your space blanket if it is cold or wet.
- **Blow your whistle** three times every few minutes, and listen for a reply.
- **Signal** by going to a clearing and waving your arms and whistling if you see or hear searchers, an aeroplane or vehicle.
- **Eat and drink** only when really hungry or thirsty, and stay away from others' supplies!

Plan properly

Every trip is unique. Draw up a Kit List with check boxes for each member of the group, plus a separate Group Kit List. Using this every time you pack, you are far less likely to end up at the end of the first day's trail without matches or fuel for the fancy new stove.

One-day outings

Personal kit

- A suitable day-pack with a raincoat, a whistle, a water bottle, some food, a warm top, and long pants.
- Preferred additional kit would include a mini-survival kit, a large survival bag or space blanket, emergency food and in colder conditions, a balaclava, gloves, and extra socks.

Group kit

- A map and compass or GPS (see p38), a first aid kit, a torch, and some extra emergency food.
- Preferred additional kit would be a Survival Pouch (see pp20–21), and in cold conditions, a small stove, fuel, a pot, mug, matches and some instant soup.

Multi-day outings

Personal kit

Equipment should relate to the anticipated weather and terrain. Take into account the 'worst case scenario' rule.

Boots

Select these with great care for quality, size and fit. Your feet are a vital part of both the enjoyment and safety on a hike. Do not compromise on boots! Buy these late in the day and after walking around – your feet will swell to the size they would be on a hike. Boots should feel comfortable while wearing two pairs

Gearing up for multi-day outings

A DURABLE LEATHER BOOTS WITH BREATHABLE SYNTHETIC UPPERS AND RUBBERIZED TOE BOX.
B BACKPACK WITH STRAPS FOR HOOKING ON EXTRA EQUIPMENT, MESH POCKETS, BASE COMPARTMENTS AND REMOVABLE DAY PACK.
C CLOSED CELL FOAM MATTRESS TO INSULATE AGAINST COLD SEEPING UP FROM THE GROUND.
D DRAWSTRING-HOOD SLEEPING BAG WITH SHOULDER COLLAR TO KEEP IN WARMTH.
E WIND-RESISTANT OUTER SHELL JACKET IN A LIGHT, BREATHABLE AND QUICK-DRYING FABRIC.
F KHAKI DESERT HAT WITH WIDE BRIM.
G FLEECE TOP WITH NECK PROTECTION.
H RUBBER-SOLED SANDALS LET THE FOOT BREATHE AND PROVIDE TRACTION.

of socks – a thin inner pair and a more cushioned outer pair. Take your own usual (high-quality!) socks when fitting boots. Sore feet, blisters and collapsing shoes can lead to disastrous delays. Treat leather boots with a softening and water-proofing agent. Wear them in before the trip. Take spare socks, to change at mid-hike, and extra laces.

Backpack

This should be appropriate to your body size, be able to hold all the items you need, and above all, be durable and comfortable. It should have a well-padded, securely fastening hip belt that helps to stabilize and hold the weight of your pack, relieving pressure on your shoulders. Avoid giving children a pack that is too large, tempting them to carry too much (your full pack should not exceed one-quarter to one-third of your body weight).

Sleeping bag

Choose according to your expected conditions – it should be light but suitably warm. Down bags are light relative to the warmth they offer, but perform poorly when wet. Synthetic fibres are bulkier and heavier, but do not lose insulating properties when wet. 'Short' bags for children are sold in many stores.

Insulating mattress

A thin, lightweight mattress provides comfort, and vital insulation from the damp and chill of the ground. Simple expanded foam Karrimat ® ones suffice, however specialist inflatable models offer enhanced comfort and insulation from seeping cold.

Raincoat

A good raincoat provides protection not only from the wet but also in cold or windy conditions. Choose one with a drawstring-type hood, and of reasonable length to cover at least the top of the legs while still allowing free movement. Those made from high-quality 'breathable' fabric keep you drier than standard plastic raincoats because they allow perspiration to pass. A good 'poncho' can be very effective.

Additional essentials

Each member should also have the following equipment:

Head lamp: a compact but powerful head lamp, spare batteries and bulbs. Make sure that it cannot switch on accidentally in your pack.

Water bottle: even in areas with good water supplies, carry a full bottle – particularly if the water you come across might have to be purified before use.

Warm clothing: do not be caught unprepared by changing weather conditions. Track pants, a warm top and a balaclava are advised for most hikes, with additional items such as gloves if cold is anticipated. Take a change of normal clothes (especially socks) in case you get wet, or for added comfort.

Other: mug, bowl, cutlery, toiletries, personal medicines, plus mini survival kit (see p20).

WHEN SELECTING A HEAD LAMP, LOOK FOR A COMFORTABLE HEAD STRAP AND A POWERFUL LIGHT BEAM WITH A LONG BURN TIME. OPTIONS FOR THE LAMP ARE TUNGSTEN, HALOGEN OR LED.

Group kit

The group should carry enough of the following items for the group's size:

Tent

Choice is dictated largely by economy and intended use. Select one that is suited to your route. Quality tents are double-skin and lightweight, and have built-in groundsheets, aluminium poles and good zips.

Compass

A good compass preferably with a luminous dial (e.g. the Silva type or Polaris compass). A Global Positioning System (GPS) device is an expensive but useful navigational tool.

Maps

A reliable topographical map (scale of 1:50,000 or more). This should be laminated or carried in a waterproof pouch.

Stove and fuel

See p69. Remember the matches and pots!

Food and water

Lightweight but nutritional food, with a fair reserve in case of emergencies. Water purification tablets and/or a water filter, to purify water before use (few water sources are unpolluted). Many newer filters are highly efficient (see p57).

First aid kit

This should contain adhesive strips, bandages, antiseptic cream, scissors, forceps, latex gloves and certain basic medicines (see p84).

Playing cards

Long periods of waiting can fray nerves and impose additional stress on a group. Games help to calm people, especially children, and take their minds off the situation.

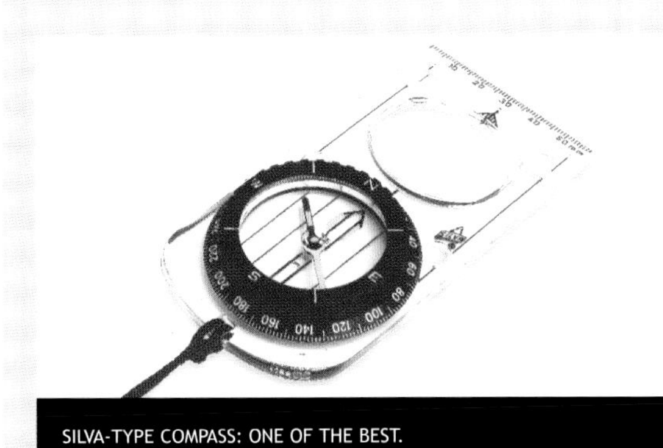

SILVA-TYPE COMPASS: ONE OF THE BEST.

A PACK OF CARDS FOR DISTRACTION.

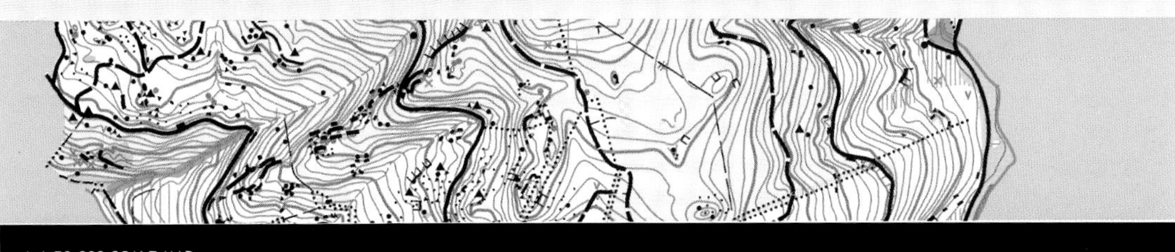

A 1:50,000-SCALE MAP.

Survival kits

Personal

The survival items in this mini kit fit into a small tin or other waterproof container that can be easily carried in a pocket or pouchbag. (It also slides easily into a jacket or briefcase pocket for Frequent Flyers!) Many hunters and fishermen like to pack the items into their many-pocketed fishing (or photographer's) vest.

Must-have essentials

- **Whistle:** the sound can attract searchers. Avoid metal whistles – they can freeze to your lips.
- **Space blanket:** compact, aluminium foil-coated.
- **Waterproof matches:** bought as such or standard matches coated with clear nail varnish or wax.

SHAVINGS FROM A MAGNESIUM BLOCK FLARE UP EASILY WHEN LIT.

Good-to-have items

- **Button compass:** should be luminous; check regularly for rust.
- **Flint:** with magnesium block attached.
- **Candle:** as light source, or to start a fire.
- **Needle and thread:** to repair clothes, etc. and for removal of splinters.
- **Safety pins:** for fastening material; also use as fish hooks.
- **Plastic bag:** needs to be strong to carry or collect water in a still.
- **Fish hooks:** for small- to medium-sized fish (e.g. gauge 5); a few pea-sized sinkers.
- **Thin wire:** e.g. brass picture wire, for snares and tasks such as fixing shoes.
- **Flexible (wire) saw:** coat with grease, keep in a plastic bag; is useful for cutting large branches.
- **Magnifying glass:** to start tinder fires.
- **Pocket knife:** small and basic, e.g. thinner Swiss Army type.

Mini medical kit

This kit should contain some basic medical supplies and tablets that are clearly labelled and packaged in plastic (check the expiry dates regularly). Suggestions are:

- Analgesic: e.g. Paracetamol (Acetaminophen in USA) and/or codeine phosphate
- Anti-diarrhoea medication: e.g. Loperamide (Immodium®)
- Antihistamine: e.g. Promethazine (Phenergan®), for insect bites, stings, allergies, etc.
- Latex gloves: worn to prevent infection of patient and first-aider
- Butterfly sutures: invaluable for holding a wound together
- Adhesive strips: preferably waterproof and in many different sizes
- Surgical blade: for cutting off dead skin, and a multitude of other uses.

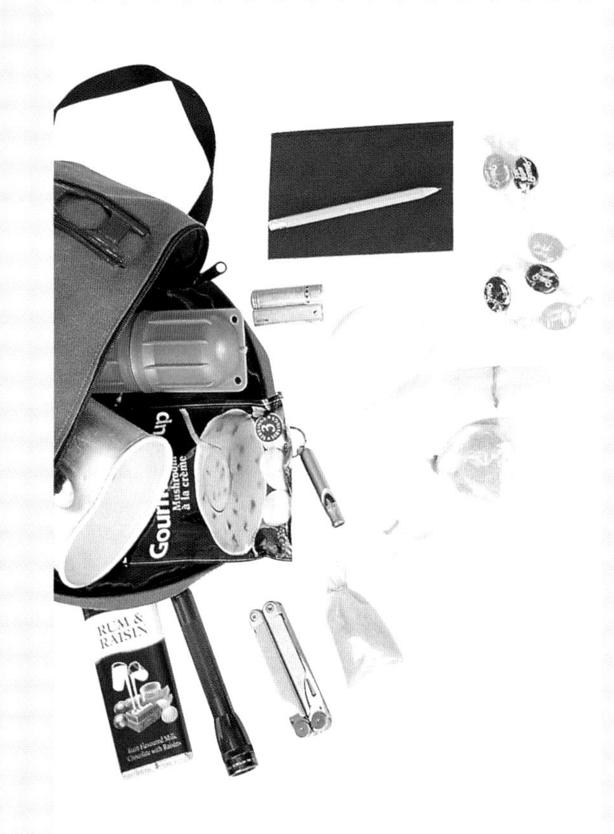

Group

Carry this kit in a single Survival Pouch (compact, waterproof and made from strong material with a solid fastening). This can easily be transferred to a vehicle, aeroplane, canoe, or hiking pack.

Must-have essentials

- **Mess tin:** to protect items; as a cooking utensil. Thin stainless steel or aluminium with a fold-away handle. If polished, can also serve as a signalling mirror.
- **Pocket knife:** the multi-bladed Swiss Army knife is good, although a strong multi-purpose tool (e.g. Leatherman®) includes pliers.
- **Solid fuel tablets:** with a small fold-away stove or potholder; can also be used as fire lighters.
- **Torch:** small, with a set of spare batteries and bulb. New compact LED (light emitting diode) models give 100-plus hours on two tiny batteries.
- **Food essentials:** sachets of tea, some sugar, milk powder and instant soup.

Good-to-have items

- **Tough space blanket:** with a reinforced nylon backing to the heat-reflective foil (red is a good colour for signalling purposes).
- **Mini flares:** particularly if heading out to sea or into remote back-country, take at least one set in waterproof containers. Check expiry date.
- **Small pocket lighter:** better than matches in wet or windy conditions. Replace lighters regularly as they can rust and leak.
- **Thin, straw-sized plastic tube:** about 50cm (18in) long; for obtaining water from various sources or as a tourniquet.
- **Energy boosters:** chocolate or any other type of candy.
- **Plastic bags:** take up any excess space with a strong plastic sheet or bag, for use as a poncho or survival bag.
- **Notebook and pencil:** for creating maps or for sending/leaving messages.

Adapting Gear to Terrain – Mountains

Personal kit

Every environment will have its own unique demands. For high mountainous areas or cold, snow-filled regions add to your kit:

- **Ice axe:** a long walking axe for moderate slopes; fasten to your belt or harness with a leash.
- **Ski poles:** on mild up- or downhills, telescopic ski poles are often more useful than an ice axe; they help to take strain off your knees.
- **Warm fleece clothing:** worn according to the principle of layering (see p52).
- **Balaclava:** reduces heat loss from the head.
- **Gloves:** many climbers wear several pairs – thin inner gloves, then thicker fleece or woollen mitts, covered by waterproof outer gloves.
- **Gaiters:** a cloth or nylon covering that helps to keep snow or rain out of the top of boots.
- **Overtrousers:** wind- and waterproof long pants or salopettes (trousers held up by shoulder straps) keep the legs dry and warm.
- **Sunglasses:** essential eye protection against glaring UV light at altitude (snow-blindness).

Group kit

This generally includes a good climbing rope and some slings, as well as specialized rock- or ice-climbing gear. Only do this level of climbing if accompanied by an expert.

CARABINERS ARE AN ESSENTIAL ITEM IN CLIMBING GEAR.

Caves

Personal kit

Pack this equipment in a small, tough slingbag.

- **Clothing:** if you intend to do any crawling, wear an overall, or sturdy trousers and a long-sleeved shirt. Tough boots or shoes are useful. As caves are seldom very cold, thin warmish clothing such as polypropylene undergarments beneath an overall are ideal.
- **Helmet:** hard protective head-covering or a reinforced cap or hat is necessary; avoid hats with large rims which restrict upward vision.
- **Lamp:** a high-quality headlamp for hands-free movement; it can be fastened to the helmet with duct tape.
- **Spare torch:** the small LED (see p18) models are ideal.
- **Batteries and bulbs:** carry at least one (preferably two) spare sets of batteries, and a spare bulb taped inside the lamp head or body.
- **Food and water:** small amounts of chocolate, energy bars and dried fruit; make sure you always take a full water bottle.

Group kit

- **Candles and matches:** for emergency light and to mark your return route; remove all traces of candle wax or matches from the cave and do not place candles where they can ruin sensitive cave formations.
- **Compass, paper and pencil:** to keep track of turns and directions taken while exploring subterranean chambers.
- **Rope:** proper 'static' caving rope, made of material less susceptible to battery chemicals than normal climbing rope, to safeguard cavers on moderate slopes or tied around ankles to back-haul in tight passages. *Note:* cave roping is highly specialized and for experts only!
- **Strong string or nylon cord:** to mark your way in complex underground passages; ensure that the string is securely tied at the start and remove it when exiting the cave.
- **First aid kit:** a basic kit able to cope with minor abrasions and scratches.

Desert

Personal kit

- **Clothing:** Lightweight cotton, loose-fitting and light-coloured to reflect heat; long-sleeves and long pants to protect against sun exposure.
- **T-shirt:** absorbs and dissipates perspiration from the skin.
- **Hat:** a broad-brimmed hat with ventilation holes for the sun.
- **Sunglasses:** a good-quality pair filters out UVA and UVB rays and prevents eye-strain in the desert's harsh reflective glare.
- **Jacket:** a warm and windproof jacket is needed, as the high desert temperatures can drop drastically after sunset.
- **Sunscreen:** a sunblock cream with a sun protection factor (SPF) of 30+.

Group kit

- **Extra water:** any vehicle travelling through a desert area (even on good roads) should carry adequate emergency rations of water.
- **Plastic sail or groundsheet:** to create shade during the day and for extra warmth at night; also useful to collect water through condensation or to make 'stills' from plants.
- **Spade:** to dig yourself out of soft sand.
- **Vehicle spares:** take tools, tyres and basic engine spares; extra fuel, as well as a fan-belt, radiator hose, fuses and other parts (e.g. spark plugs, points and a condenser) for remote areas.
- **Food:** carry provisions that will last at least three days.

Jungle and tropical areas

Personal kit

- **Clothing:** essentially as for desert conditions, but tougher material preferable.
- **Drawstring trousers:** to keep out leeches and other crawling creatures (elastic bands can also be used).
- **Mosquito-net fringe on hat:** vital – this can be fitted as needed (evenings or while resting).
- **Insect repellent:** essential – apply to hands, arms and other exposed body parts, avoiding forehead and around eyes.
- **Anti-malarial medication:** have your doctor or clinic advise you before you visit any affected regions.
- **Mosquito net:** suspend over your bed at night; most effective mosquito protection.

Group kit

- **Medical kit:** injuries can rapidly become infected in the tropics; any wounds should be covered with a sterile, waterproof dressing; carry anti-fungal cream for the feet.

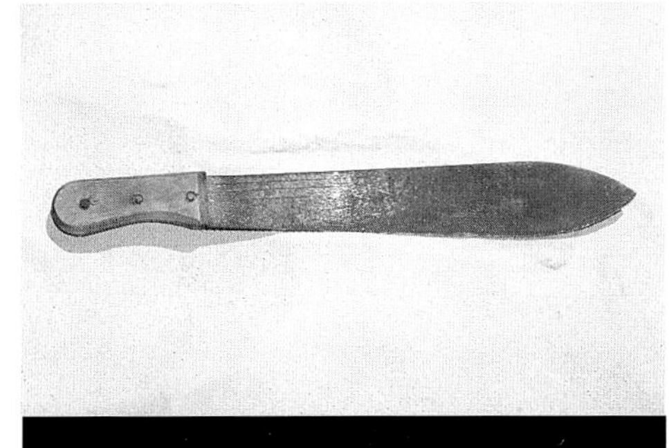

ALSO INCLUDE A MACHETE OR A LARGE KNIFE IN YOUR GROUP KIT – IT IS USEFUL FOR CUTTING AWAY VEGETATION.

Sea voyages

Ocean-going vessels are required by law to carry certain basic rescue and emergency gear, which includes life rafts (or life jackets on smaller craft), radio, flares, and often rescue beacons.

Personal kit

- **Warm, waterproof jackets:** on a longish sea trip, cold can pose a serious danger.
- **Life jackets:** sufficient for the crew; all should know how to use them.
- **Strobe lights:** powerful and compact, easily visible at night, can be fastened onto life jackets; these flash for between eight and 48 hours, depending on battery size and status – check and replace batteries regularly.

Group kit

- **Life rafts:** usually easily inflatable, should be made of tough canvas or rubber, and must be stable, have plenty of places to hold onto and preferably have some form of built-in shelter; there should be basic survival gear (including food and water) on board.
- **Survival kit:** (see pp20–21) keep in an easily accessible pouch.
- **Flares:** both parachute flares and red handheld or smoke flares assist your chances of being seen. Check expiry dates.
- **Waterproof containers:** to hold vital personal effects, as well as emergency rations, waterproof matches, maps and flares.
- **Sea anchor (drogue):** this can be streamed behind the boat to keep the bow facing into the direction of the weather and thus restrict drifting.
- **Gaff and net:** for catching fish; protect the gaff point in cork or similar material.
- **Cord:** strong nylon cord has many uses in shipboard survival.

THE WEARING OF A LIFE JACKET IS REQUIRED BY LAW.

Working with basic tools

Tools to buy

Take these along if you are venturing into jungles, or serious backwoods. It is wise to keep them in your vehicle if you travel into remote areas.

- **Sheath knife:** choose one that has a solid tang going right through the handle, as this can be used even if the wooden or plastic handle breaks; also a sheath to hold it securely, with a belt loop. Some good-quality foldable knives are also excellent.
- **Kukri, parang or machete:** hybrids between a knife and an axe; use for hunting, cutting wood and clearing away vegetation.
- **Axe:** one of the most valuable tools in the bush – handle with care.

AN ALL-PURPOSE, FOLD-UP KNIFE IS HANDY IN THE WILD.

Making tools

Tools can be fashioned from diverse sources including glass, metal or the tough plastic parts of vehicles. Some suggested tools are given below, but in all survival situations, improvization is the key.

Stone tools

It is possible to split flakes off a solid piece of rock with the help of smaller stones, to make tools such as those used by prehistoric man. The finer flakes can be created from hard wood or bone. Such tools can serve as axe heads, knives, scrapers, or spear points.

Heavy, smooth black rock – for example, obsidian or flint – makes good tools.

Wooden tools

A wooden spear is a simple but effective tool or weapon; harden the point with a few seconds of repeated heating and cooling in a fire, rotating the point while you do this.

Glass knife

Use a shard of glass wrapped in leather or cloth to gut fish or other animals. To make one, wrap a glass bottle thoroughly in cloth, then tap firmly on a hard surface. Open the cloth and select a long piece with a sharp edge and thick base for a handle. A robust covering tied over the handle section is essential to prevent injury to the user.

Bamboo knife

This is a very useful tool for hunting smaller animals, or as implement for digging out roots and bulbs. Cut the end of a long section of bamboo at an angle to create a point. If you have no knife, break the bamboo by snapping it over a tree branch until you get a suitable point, then rub the edges along a rough surface to further sharpen the point.

Making a Move

In a disaster situation, movement should be seen as the last resort unless there are pressing reasons not to stay, there is a clear-cut route to help, or you are sure that rescuers are unlikely to find you easily in your present location.

Move or stay?

Consider the following before making your decision:

- If a vehicle is involved consider staying close to it as it may be easier to spot than a group or an individual on foot.
- Is anyone likely to know your present location? Does anyone have details of your proposed route? Have you moved far off your itinerary?
- Do you know where you are relative to your original route, a road or some form of civilization? Would you simply be heading off blindly?
- Is the nature of the terrain suitable to the skills of the group? Are the potential hazards such that movement is an unwise option?
- If there are injured group members, how serious are their injuries? Should they be moved? How easily can they be moved?
- What is your food, water and equipment stock? How long will these last?
- Do you or your group have any special needs, e.g. special medication (for chronic conditions such as diabetes or hypertension)?
- Is splitting the group a viable option? Would it be feasible to send some members for help and leave the rest of the group behind?

opposite THERE IS A FINE BALANCE BETWEEN DECIDING TO WAIT FOR POSSIBLE RESCUE OR RESOLVING TO KEEP MOVING AND RISK THE ENDANGERMENT OF ANY MEMBER OF THE GROUP.

Advantages of staying

- Movement requires energy; staying allows your food supplies to last longer.
- You can make a more permanent shelter.
- The risks are known; unfamiliar terrain may expose you to greater hazards.
- You can plan and implement signals to alert rescuers, such as making smoke.
- Injured, old or sick members of the group will not be exposed to the stress or rigours of a move.

Advantages of moving

- Good shelter, food and water might be easier to obtain elsewhere.
- If rescue is unlikely for some time and resources are insufficient, then movement can save the day.
- If camp is made in one area for an extended period, hygiene and sanitary conditions might deteriorate.
- If at high altitude, it might be necessary to descend to maintain the health of the group.
- It might be psychologically better for the group to take action rather than to wait things out.

Setting out

Make sure that you have taken everything that might be of use – strip the vehicle. Ensure that all group members know the intended route in case someone gets lost. Leave some clear form of message for potential rescuers, indicating your direction of travel and time of departure.

If you have no map and no clue as to your exact whereabouts, then the best option is often to follow a watercourse. Most rivers eventually lead to lakes or the sea and are likely to have settlements nearby. Try to follow the general direction of the river according to the lie of the land. When the river widens it can offer a transport mode on a raft.

If you are moving with a group, keep together. This is difficult with large differences in age or fitness, or with injured members – and even more so in fog, falling snow, rain or mist, or when moving at night. It calls for organization and discipline – not always easy to impose!

Have a path-finding scout (or two) slightly ahead of the rest, while members of the main group follow the leader. Place a responsible 'tail-end-Charlie' to ensure that no stragglers fall behind. Scouts should not lose visual or auditory touch with the group, and each group member should keep in constant contact with the person immediately behind and in front of him/her.

Stop for regular checks and head counts. Every member should know what action to take in the event of being separated from the group. In general, this would entail stopping and waiting while calling occasionally. If each group member or subgroup has a whistle, it improves the chances of locating someone.

A small group of faster, fit members can act as a scouting party to find food, water and shelter, or cut a path for the main group. However, they must mark their route well, by cutting notches into trees, tying knots in grass, placing sticks or stones in patterns on the ground or tying pieces of material to prominent natural features i.e. large rocks or trees. Remember that what you might think is an obvious marker might not be so clear to those who are following, especially if it is raining or snowing. Move back, and then up to your marker(s) to check whether they really are clearly visible.

On difficult terrain, it is vital to constantly be aware of whether any of the group members are struggling. On steep slopes, join hands with a firm hand-to-wrist grip because visibility is restricted and it is not always possible to judge the size of drop-offs (cliffs or sudden steep slopes) correctly.

WHEN BAD VISIBILITY HAMPERS PROGRESS, BECOMING SEPARATED IS A REAL THREAT. THIS CAN BE SAFEGUARDED BY LOOPING A ROPE AROUND THE WAIST OF THE LEADING PERSON AND LINKING IT TO ONE OR MORE COMPANIONS, ENSURING THEY STAY TOGETHER.

Making progress easier

Makeshift frames, backpacks and sleds can be made using branches and cord (see illustrations). If your move is expected to last long or entails a reasonable distance, with injured people or many bulky supplies to haul, make a carrying frame or sledge. The two-toed sled (travois) is the easiest and most basic to create, while the curved runner sledge is much easier to move than the simple sled, especially on snow, ice or smooth ground.

Makeshift backpack

Branches can be lashed together to make a frame; gear is wrapped into a groundsheet or large item of clothing and tied to the frame. Wrap clothing around the frame as padding for back and hips. For comfort, makeshift shoulder straps should ideally be as broad as possible.

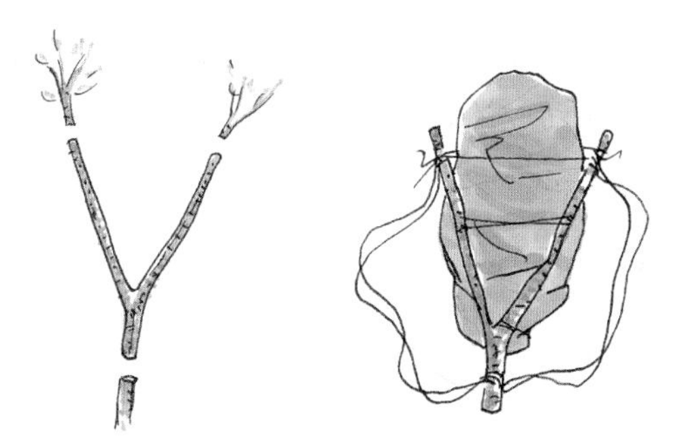

Hudson Bay pack

A large square cloth or oilskin is most suitable here. Tie a stone into two diagonally opposite corners, using strong cord or bark, to enable sturdy fastening. Roll up your goods well, lengthwise, then attach cord between the two stone-carrying corners, and hoist across your back or around your waist.

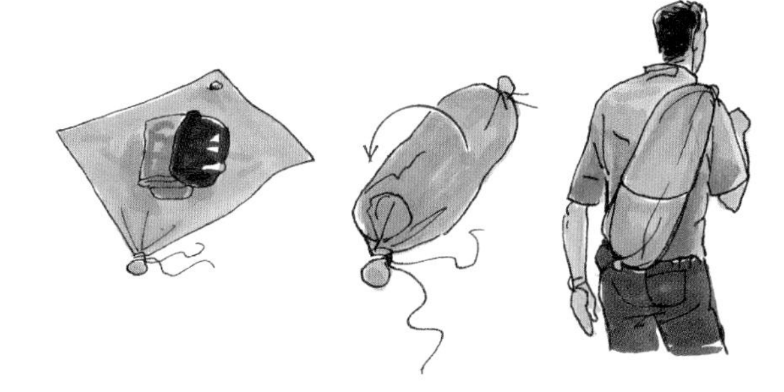

Two-toed sled

Use a pair of long branches or even two pack frames to create parallel runners that drag on the ground. Runners can be joined by passing them through the sleeves of a shirt/jacket, or lashing on cross-branches.

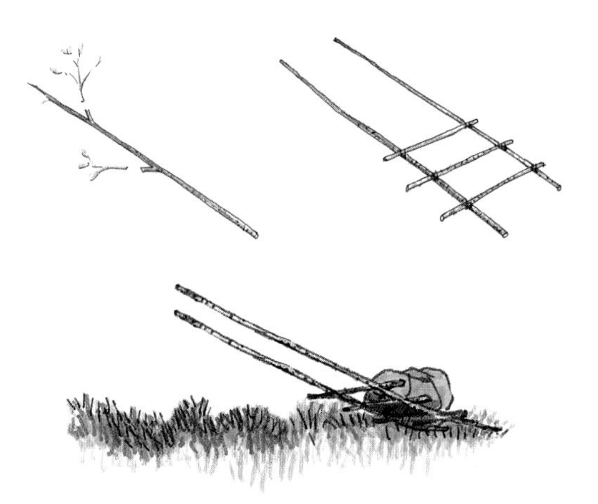

Curved runner sled

Create the curve at each end of this sled by bending up the main runners; fix this curve by bracing the lifted end to a horizontal crossbar on the base of the sled by means of a sturdy branch, which is lashed to the crossbar with cord. The sled can be braked down a steep incline by pulling on a rope attached to the back.

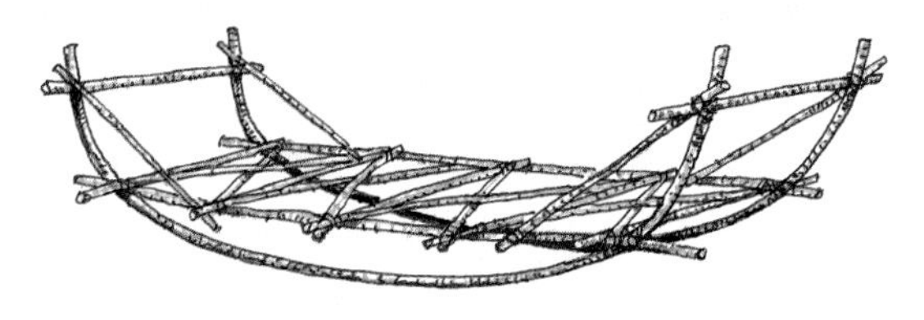

Progress at night

This should only be undertaken in the case of medical emergencies, in desert areas where it is inadvisable to travel during the day, and perhaps on snow or glaciers, which become safer when they harden at night.

Using torches makes immediate movement safer and easier, but you lose distance vision, and it is harder to keep track of position. Sit and wait a while before starting out as night vision takes over half an hour to develop fully. This important adaptation of your eyes can be destroyed in an instant by any bright light – even a flaring match. If you need to use a light to read a map or for any urgent reason, make the group keep their eyes shut to keep their night vision intact.

Disadvantages of night travel include the inability to anticipate obstacles and steep drop-offs and the usual difficulties inherent in making sure that the group does not split up.

Sticking to the route

In the dark, one way of maintaining direction is to use the Triplet Method (illustration below left). Establish a visible line of travel by positioning group members and leader in one spot, an 'aimer' halfway along the line and a 'target' at an equal (visible) distance further along. When the group has reached the aimer, he moves off beyond the target to become the new target.

ALTHOUGH ONE'S VISION DOES ADAPT TO CONDITIONS OF DARKNESS, JUDGING DEPTH IN UNDULATING TERRAIN BECOMES DIFFICULT.

KEEP AS MANY OF YOUR CLOTHES AS DRY AS POSSIBLE WHEN CROSSING WATER, AS HYPOTHERMIA CAN OCCUR VERY QUICKLY.

Crossing rivers

Rivers can be a real problem for a person or group, especially if members are tired, cold and disoriented. Apart from death or injuries, getting wet can result in hypothermia (see pp80–81), even in seemingly warm conditions. Before crossing, remove most of your clothes. Hold them above your head or wrap them in waterproof material or a container. Shoes should be kept dry; wear only if the riverbed is rocky or there may be hidden obstructions underwater. If you only have one pair of socks, wear shoes without them.

If the river seems to be flowing quickly but is likely to subside, it is often wisest to wait it out or to look for an alternative route. If there is no option, then spend some time moving up and down the bank, studying the river and its flow patterns in order to choose the best crossing point.

After any river crossing, try to get group members as warm and dry as possible and watch carefully for symptoms of hypothermia.

Ropeless river crossing

- Your method will depend on the width, depth and power of the river and the group's strengths and weaknesses.
- Choose a crossing spot with no visible hazards below it; don't cross just above cataracts and waterfalls. Avoid crossing on river bends – water always runs fastest on the outside of bends. Cross diagonally with the current; don't try to fight it.
- Face upstream so you can avoid any debris sweeping towards you. Large boulders in fast-moving water can form dangerous eddies and whirlpools above and below them.
- Crossing in a group line with a sturdy stick or pole held horizontally across the chests gives support to smaller, weaker or less sure-footed individuals. Place the strongest person upstream to break the flow of current, or form a group 'huddle' with all facing inwards, link arms across each other's shoulders and cross by shuffling sideways. The person on the leading side uses a stick for balance as well as to assess the depth and risk of obstacles on the riverbed. For an individual wading across wide, shallow, slow-flowing streams, use a pole or stick both for support and to test the depth ahead.
- If crossing with a heavy backpack, loosen the hip belt of the pack. If you lose your balance you can remove the backpack to avoid being swept along face-down underneath it. Use flotation aids to tow weaker members or kit across larger rivers.

Rope-assisted crossing

The safest method is a continuous loop (see p34). Never try to pull a person who is swept away back upstream as he/she could be drowned by the strong force of the water. Allow him/her to be swept sideways to the bank on the tensioned rope. A safety line can also be tied across the river and the group can cross holding on to it while being belayed by another rope, which should be angled slightly downstream.

Rope-assisted crossing

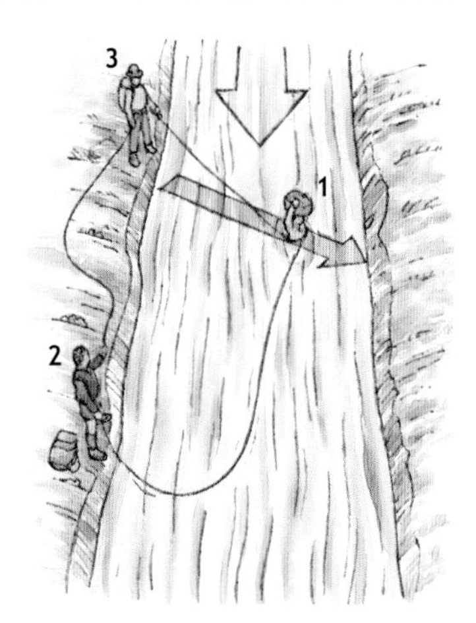

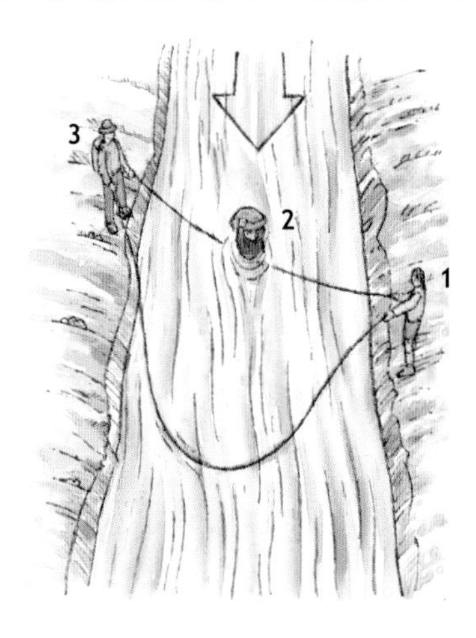

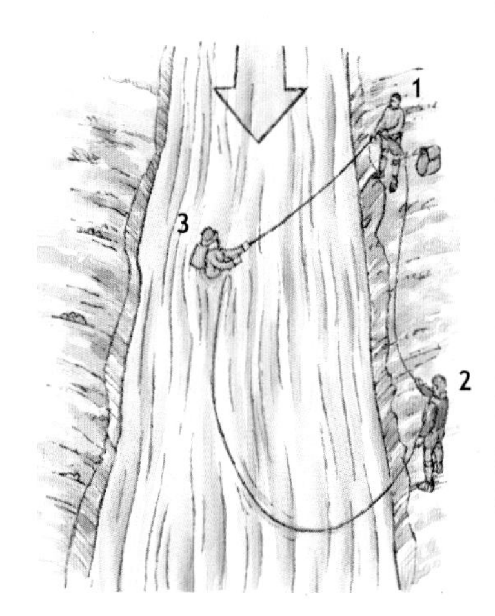

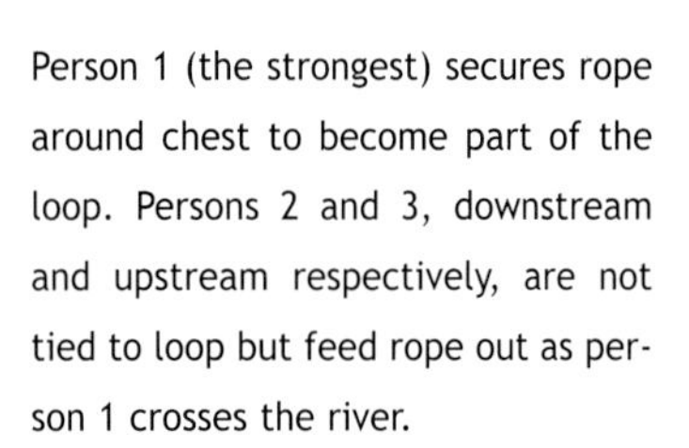

Person 1 (the strongest) secures rope around chest to become part of the loop. Persons 2 and 3, downstream and upstream respectively, are not tied to loop but feed rope out as person 1 crosses the river.

Person 1 reaches the bank and unties himself. Person 2 secures himself to loop before walking to 3. He enters water and is belayed (rope held taut) by 3, while 1 keeps rope taut from opposite bank to guide 2 across.

Person 2 reaches person 1 on the bank, unties and walks downstream to take up loop. Person 3 secures himself and crosses, supported by person 1 who holds rope taut and bears most of rope weight, while 2 belays.

HANDRAILS AND SIDE SUPPORTS ON A MONKEY BRIDGE MAKE THE PRECARIOUS CROSSING OF THE CHASM MUCH SAFER.

Makeshift bridges

If you need to cross a stream regularly in a long-term survival situation or near a base camp, it is a worthwhile exercise to build a bridge.

Monkey bridge

You need a person or team on both sides of the gap. Each must create, for each end of the bridge, a simple, well-anchored X-frame around 2m (6½ft) high and thoroughly shear-lashed (see p71) in the centre. Secure a crosspiece low across the legs of each X-frame, dig the frame's legs into the ground, then angle the frame to about 45 degrees and anchor the crosspiece to trees or solid stakes. A rope is securely tied to a tree and passed over both X's, then fixed again, to serve as the walking rope.

Placing material under the rope at the X prevents it wearing through under the weight of walkers. Raise the X-frames vertically to fully tension the rope, then attach two handrail ropes to the upper arms of each X, securely anchoring them at each end.

Simple log bridge

This bridge can be constructed from one side of a river. Start by securing a short log on the bank edge (a) at cross-angles to the intended bridge, using stakes to fix it in place. Brace a long, bridging log against this (b), using rope to raise it vertically alongside the bank (c), then swing it across the river and drop the end onto the opposite bank (d). Slide a second (e), then a third, log across the first, manoeuvring them into place (f). Secure with metal or wooden stakes on each side.

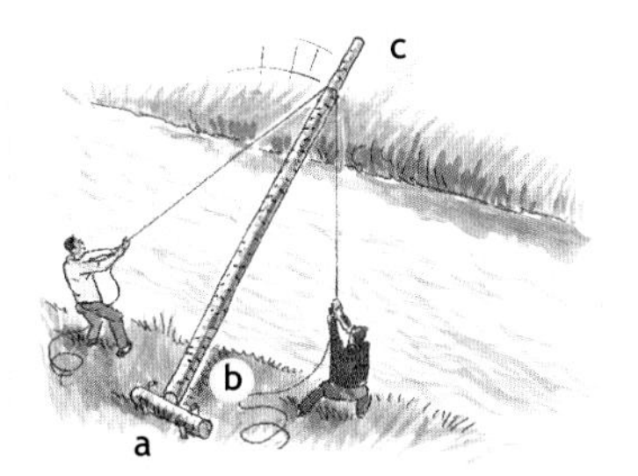

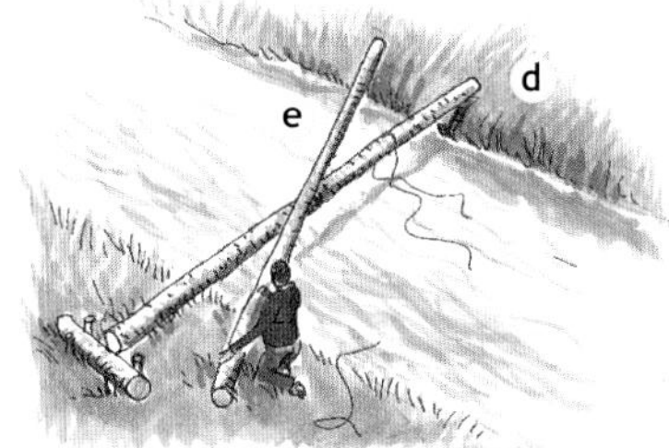

Rafts

A raft can be used to ferry small children, aged or injured members, as well as supplies across a wide, slow river. Rafts can be made from logs or from thick bamboo, or even thick canvas on a frame. Take into account the intended load and make your craft large enough to cope with it. The raft will probably be heavier than you anticipate, so build it close to the bank.

Begin with a base of two long poles lying across a shorter pole (a). Then cut a notch along the length of two logs, to act as retaining logs (b). These are placed, notch face-up, at right angles at either end of the raft base (c); the rest of the logs are arranged to lie firmly within the notches (d). Carve a notch along the length of two final logs, and secure them crosswise at each end of the raft (e), lashing tightly to hold the raft together.

Oars can be made from a flat piece of wood or several equal lengths of bamboo lashed together, then tied onto a long branch. A small X-frame (f) lashed to the back of the raft can act as a support for the oar or rudder (g).

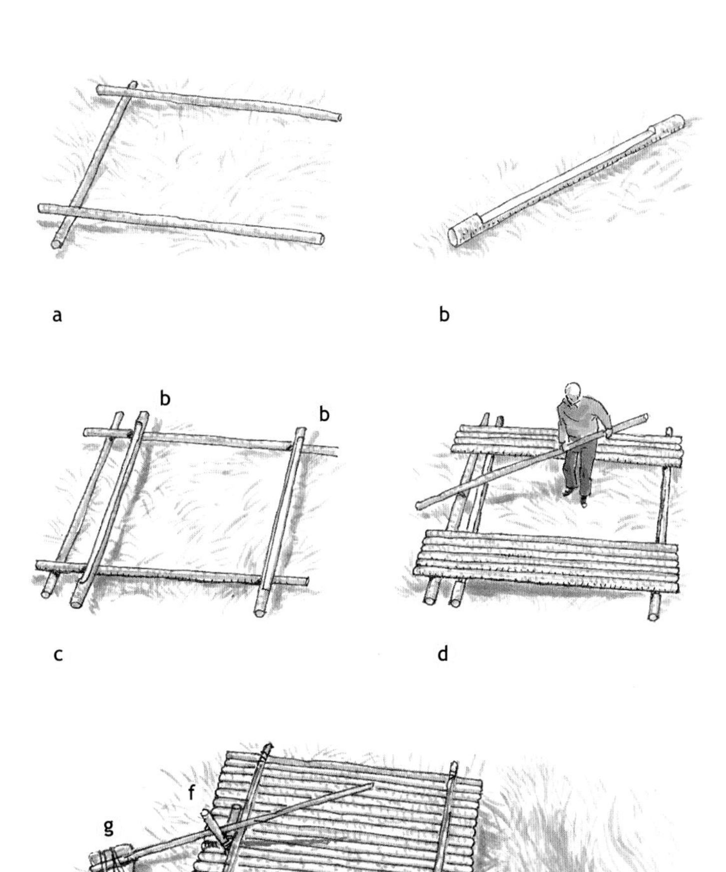

Navigation

Finding your way in unknown territory is best achieved with a map, and a compass or GPS. If you have neither of the latter, you will have to rely on natural means of finding direction (see p40–41).

Using maps

Maps indicate the following:

- **Direction via north–south lines** (or grid lines). Once a map is orientated correctly, natural features on the map such as rivers, hills and valleys should match what you see on the ground.
- **Distance via the scale of the map.** Maps have different scales and thus varying degrees of accuracy. A hiking map usually has a scale of 1:25,000 up to 1:50,000. At a scale of 1:50,000, for every 1cm you measure on the map, there are 50,000cm in reality on the ground – 1cm represents 500m. (Alternatively, 1in represents 50,000in.)
- **Heights, via contour lines.** These also give an indication of the slope or gradient and spot heights (e.g. peak heights). The contour interval is represented on the map by the vertical height difference between each line, usually about 10–20m (40–80ft) in reality. The closer together the contour lines, the steeper the slope. Take careful note of slope angles. A cliff or steep drop-off is indicated by many contour lines that are very close together.
- **Special features and geographical landmarks** i.e. lakes, rivers, cliffs, roads, buildings and vegetation.

Matching map to area

A map is only of value if you can orientate it to your position and surrounding terrain. The first objective, therefore, is to match up your map with north and your environment.

Map vs. environment

Orientation via terrain: if you know roughly where you are and visibility is good, turn the map until visible features such as hills, peaks, rivers or lakes, represented by contours and other means, match the terrain you are in.

Orientation via compass: it is advisable to do this anyway, since lining up via the terrain could be less than accurate! The most suitable compass is the Silva type, which is fairly inexpensive, robust and easy to use.

Navigating with the map only: once the map is orientated, you can use it to establish your exact position and to plot your route, e.g. 'down the valley on our east, over the neck, down the southwards ridge to the road'.

Follow your selected route using the map to guide you by matching the features on the map to your physical environment.

Below In terms of map-reading, you need to correlate physical-feature map symbols with the environment around you.

Using map & compass or GPS

Compasses

A compass is a great aid to route finding if you know how – using a compass accurately with a map *can* get tricky at times!

Magnetic north and true north The Magnetic North Pole is not exactly at the earth's geographical, or True, North Pole. This difference is known as magnetic variation (or magnetic declination). True North differs from Magnetic North (or 'compass north') by a varying number of degrees according to where you are on the earth. The map should somewhere give an example such as: magnetic variation (or declination) 15 degrees east. This means that the magnetic pole actually lies 15 degrees east (to the right) of true north in the area covered by that map, so the compass reading is off by 15 degrees east of true north.

Rotate the map 15 degrees westwards (left, or anticlockwise) from the compass north (magnetic north) to line up on true north, so it matches up exactly with the terrain as you see it (see diagram below).

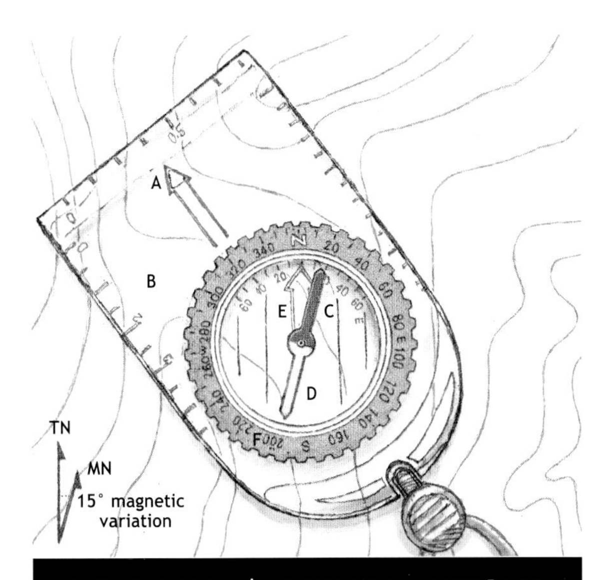

THE SILVA TYPE COMPASS – **A**: DIRECTION OF TRAVEL ARROW; **B**: BASE PLATE; **C**: RED MAGNETIC (NORTH) NEEDLE; **D**: WHITE (SOUTH) NEEDLE; **E**: RED NORTH (0°) INDICATOR; **F**: ROTATING BEZEL.

PROFICIENCY IN MAP-READING SKILLS WILL STAND YOU IN GOOD STEAD WHEN TRAVELLING THROUGH UNFAMILIAR TERRITORY.

Improvised compass

- A compass needle can be created from a sliver of iron or steel, or a piece of razor blade.
- A good battery (from a car or a series of lamp cells) and a coil of wire can give a strong enough current to magnetize steel. The harder the metal (e.g. a tempered steel needle) and the tighter the coil, the longer the magnetism will last. Make and break contact quite a few times to strengthen the magnetism.
- Repeatedly stroke a needle in one direction on silk (or even certain synthetic fabrics). Re-magnetize regularly as the magnetism in your makeshift compass needle diminishes quite rapidly.
- Magnets are found in car generators and alternators as well as in radio loudspeakers and toy motors. If you have a bar magnet, you can use it directly as a north indicator. A ring magnet can be used to magnetize a smaller piece of metal.
- Allow the magnetized metal to swing freely: suspend it from a very light thread or float on water. As the improvised needle indicates only a north–south line with no indication on which is north, it should be used together with a map or natural way of finding north to determine this.

Global positioning systems

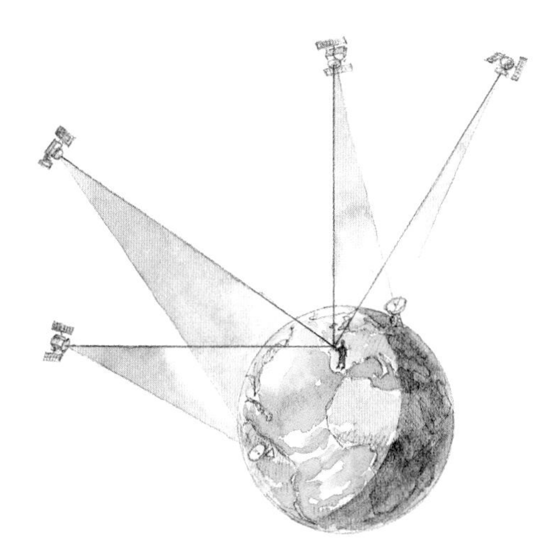

The Global Positioning System (GPS) uses over two dozen permanently orbiting satellites to establish the location of your GPS receiver via the triangulation of up-and-down radio signals. The coordinates of your position (where you are on the surface of the earth in terms of latitude – east and west – and longitude – north and south) will appear on the GPS as a set of six to 14 figures, known as the grid reference. For example: 'E40°45.28; N35°55.42' means your absolute position is 40 degrees, 45 minutes and 28 seconds east, and 35 degrees, 55 minutes and 42 seconds north. This is usually accompanied by an altitude reading, e.g. 1045m (or 3428ft).

Most GPSs will indicate which way you have to move to reach your base or a point you have chosen from map coordinates. Some GPS devices also give altitude and barometric trends (for weather prediction), act as a compass, plot the course you have travelled and supply positional information for rescuers.

Perhaps their biggest limitation is battery power – when the batteries run out, they are of no further use. For this reason, wise explorers take an 'old-fashioned' compass as a backup!

TECHNOLOGY IN THE FORM OF GPS HAS GREATLY IMPROVED TRAVEL FOR MODERN ADVENTURERS – THE TRIANGULATION OF SIGNALS ACTIVATED BY SATELLITE INSTANTLY ESTABLISHES YOUR PRECISE LOCATION IN THREE DIMENSIONS, LATITUDE, LONGITUDE AND ALTITUDE, WHICH IS CRUCIAL IN SURVIVAL SITUATIONS.

Map & compass

At night or in poor weather conditions you may have to depend on a combination of map and compass by which to travel. The compass bearing will give you the direction. Bearings run from 0° (north) via 90° (east) through 180° (south) and 270° (west), with the full circle being at 360° (north again).

A First establish precisely where you are on the map – your present position (X) then select your destination on the map (Y). Place the compass so that the edge of the base plate (and the direction of the travel arrow) run from X to your destination Y.

B Rotate the bezel so that the north indicator on the bezel lines up with the north–south lines (pointing to true north) on the map. Read the figures where the base line of the direction of travel arrow meets the bezel. This is known as your true bearing.

Convert this to magnetic (compass) bearing by adding or subtracting (as given on the map). If the variation is given as: 12° west, add this to the true bearing. If it is given as: 12° east, subtract it from the bearing; e.g. for a declination of 12° W on a map (true) bearing of 38°, add 12° declination to give a final 50° magnetic (compass) bearing. Set this last bearing by turning the bezel to 50° to line up with the base of the travel arrow on the compass.

C Hold the compass level so that the needle swings freely. Rotate the base plate until the red north point lines up with the north indicator on the bezel. The direction-of-travel arrow now points in the direction you must head (i.e. true bearing 38°, plus 12° W magnetic variation, to give an adjusted direction of 50°).

In brief, the magnetic compass needle always points to Magnetic North – and you have just set on your compass, using the direction-of-travel arrow, the way you should be heading. Provided you keep the magnetic needle lined up with the north indicator arrow on the bezel, and follow the direction-of-travel arrow, you will safely be heading towards your destination.

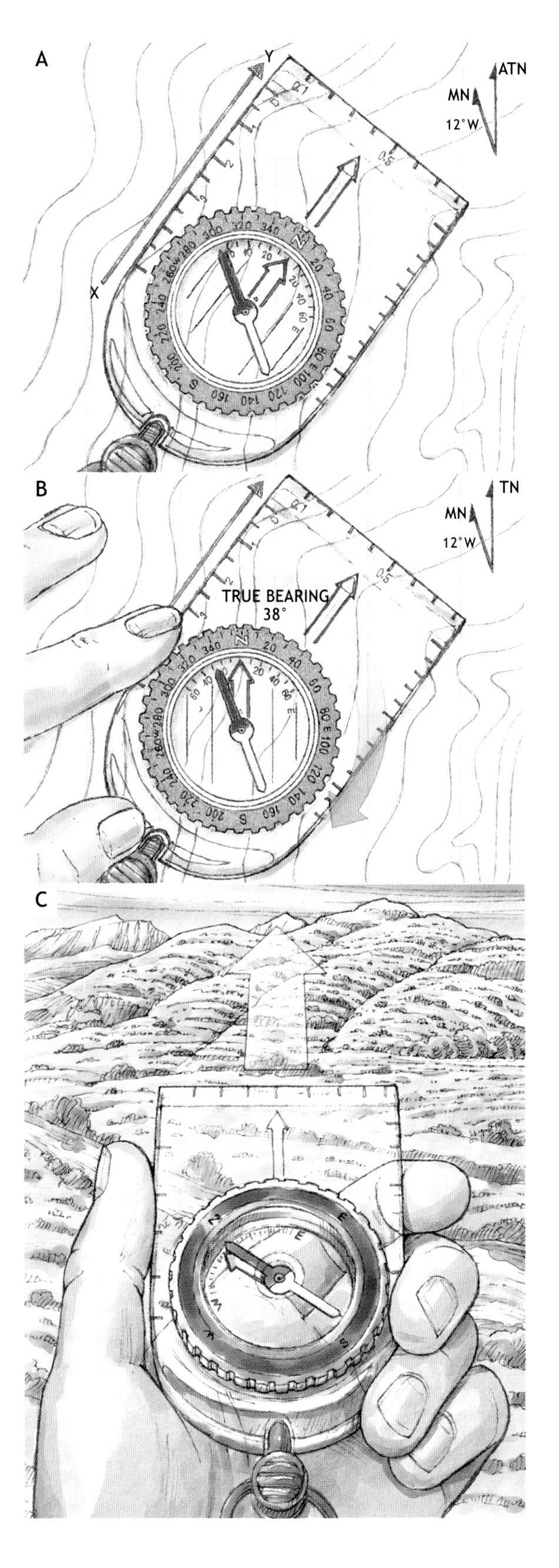

Finding direction without a compass

The sun

The sun rises roughly in the east and sets roughly in the west. There is quite a bit of seasonal variation due to the tilt of the earth on its axis. If you are in the Northern Hemisphere, the sun will be due south when at its highest point in the sky, while the opposite is true for the Southern Hemisphere.

A Shadow Stick can be used to find both your direction and the time of day.

Place a tall vertical stick in the ground. Make a mark at the tip of the shadow (West). Wait at least 15 minutes before making a new mark at the shadow tip (East). Join the two for an East–West line. North–South will be at right angles to this.

The shadow moves clockwise in the Northern Hemisphere and anti-clockwise in the Southern Hemisphere; the illustration shows a Northern Hemisphere reading.

The stars

Stars appear to 'move' across the sky because of the rotation of the earth. As the north, or pole, star is at the axis of rotation, it 'stands still' and can thus be used to find north in the Northern Hemisphere. Use the distinctive Big Dipper and the Milky Way to identify the pole star.

In the Southern Hemisphere, use the conspicuous Southern Cross and its two pointers. Extend an imaginary line through the long axis of the cross. Next, take a line that bisects the pointers and extend it, and where these two imaginary lines meet is the Celestial South Pole. Drop a line straight down to the horizon to give true south.

In the Northern Hemisphere, the 'stationary' pole star (Polaris) – in the constellation known as Ursa Minor (the Lesser Bear) – is located by extending a line from the two bright stars that form the left-hand side of the Big Dipper's bowl (in the constellation Ursa Major, or the Great Bear). Above 60–65° north latitude, Polaris is too high in the sky to be an accurate guide to north.

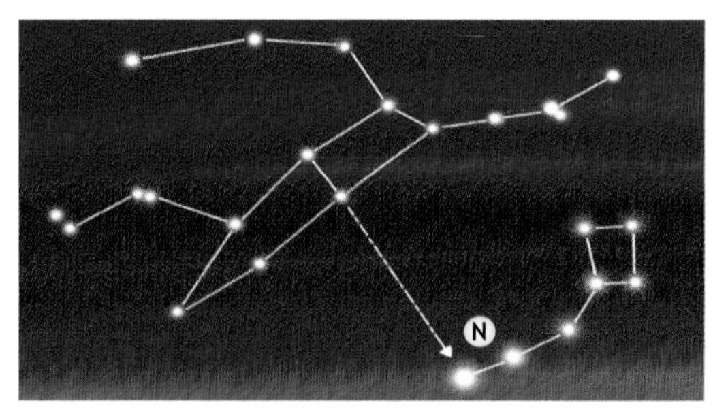

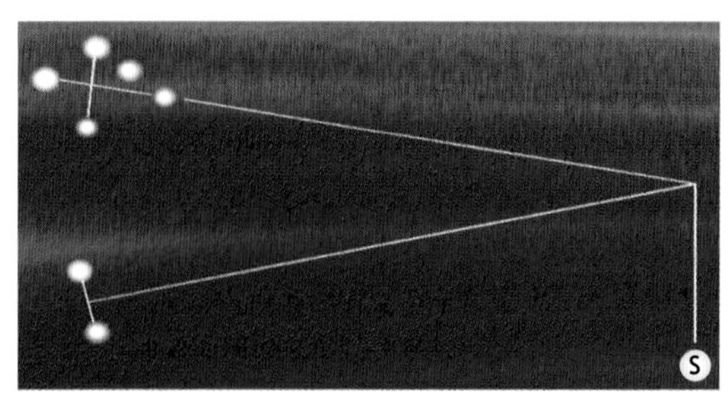

Finding direction via Nature

Lichens and moss: These grow in greatest profusion on the shaded side of trees (right) – in the Southern Hemisphere, the south side, and the north side in the Northern hemisphere.

Trees: In older trees, the annual growth rings (which can be seen on a tree stump) are larger on the side that faces the Equator.

Flowers: These mostly face the main course of the sun to flower better (above left) – north in the Southern Hemisphere and vice versa.

Birds and insects: The African Weaver bird builds its nest on the west side of trees. Termites build their nests orientated north to south to maximize shade in the heat of day (left).

River flow: If you know the principal direction in which a major river flows in the area, this would help to confirm your direction. Bear in mind that many rivers meander.

Communication and Rescue

Being seen and rescued is doubtless one of your top survival priorities. Once your ETA (estimated time of arrival) is well past, it is likely that some form of search will be organized, especially if you have left details with friends or family members. If you are not expected for a while, things might take a little longer!

Persuading the authorities to organize and maintain an effective search is not always easy, especially in Third World countries. The persistent presence of friends and family members and strong political or economic pressure may be necessary to keep organized search attempts active. It is worth priming your friends and family about this 'just in case'.

Signalling

Signalling can range from the oldest and simplest – mirrors, whistles and fires – to the most modern – radios, cellphones and transponders. Be imaginative in making visible signs and signals from whatever you have at your disposal, be it pieces of vehicle wreckage or natural and other materials.

above ANY FLATTISH SHINY OBJECT CAN BE USED TO REFLECT THE SUN.
opposite RED FLARES ARE THE MOST VISIBLE, PARTICULARLY AT NIGHT.

Traditional options

Mirrors

Over 80 per cent of air search finds have been in response to mirror signals. If you don't have one, use a flat shiny metal object or glass bottle bases.

A sighting hole in the middle helps you aim it at a rescue vehicle (i.e. a ship or aeroplane). If there is none, place your free hand in front of the mirror so that it partly hides the aeroplane or ship, shift the mirror until the sun's reflection is on your hand, then remove your hand. Move the mirror in slow arcs to make sure the flash is seen.

Signal fires

Pre-arrange one or more stacks of firewood so they can be lit swiftly when the need arises. If there is enough fuel available, keep a fire burning permanently. Fires arranged in a triangle are a universal distress signal. Alternatively, by chopping green branches or car tyres into small pieces you will quickly be able to produce thick black smoke – pick your site for maximum visibility. Fires made on small rafts in the middle of streams or ponds in an area with few clearings have also proven effective in the past.

Take care that your fire does not get out of control, and become a hazard to yourself and others.

Signs

Material that contrasts with the background makes ground-to-air signalling more visible (the larger, the better). Make a cross, triangle or an SOS pattern (Save Our Souls . . . - - - . . .).

Successful methods include controlled burning of bush patches, laying out logs and stones in a pattern, and even diverting a small stream to make ponds that resemble an SOS signal.

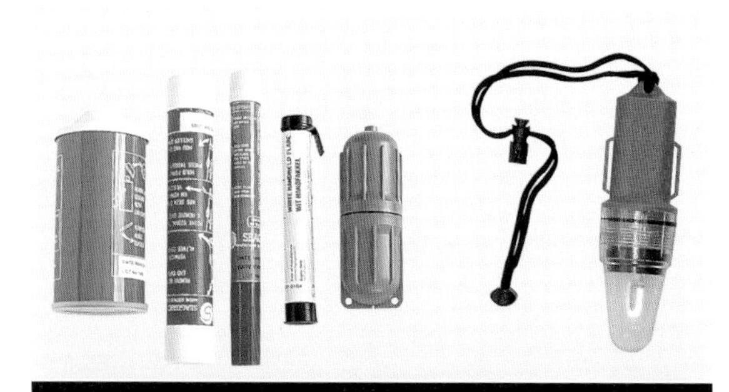

A SELECTION OF RESCUE FLARES AND EQUIPMENT (FROM LEFT TO RIGHT): SMOKE SIGNAL FLARE; DISTRESS ROCKET; HAND-HELD MARKER FLARE; WHITE HAND-HELD FLARE; MINI FLARES IN A WATERTIGHT CONTAINER; AND A WATERPROOF FLASHING STROBE LIGHT. THE LATTER IS VERY EFFECTIVE FOR SIGNALLING AS IT IS SUFFICIENTLY BRIGHT TO ATTRACT ATTENTION OVER LONG DISTANCES.

Light

Torchlight is highly visible at night for signalling from mountaintops, cliff faces, boats and islands. Three repeated flashes or the SOS signal are universally acknowledged distress signals. (The International Mountain Distress Signal is: six flashes a minute, wait a minute, then repeat.)

Flares are by far the most effective signalling devices. Red is the most visible. Use flares only when you are almost certain they will be noticed. Remember searchers will probably be looking in a 360° arc – fire several flares in quick succession to ensure they see you. All hand-held flares should be raised above the head, well away from the face on the downwind side of your body and craft. Flares become hot during use, so take care not to burn yourself or start a fire – especially if you are in a rubber or wooden craft.

How to assist rescuers

If you are part of a group that is lost, assist searchers by leaving clues of your whereabouts and movements: build cairns, make ground markers with your initials and an arrow to show direction of movement, cut arrows into trees or bend twigs. If you are sheltering somewhere such as a deep cave, leave markers at the entrance in case you fall asleep or become unconscious. Should you be snowbound, these markers can be covered by fresh snowfall, so arrange them at a suitable height. If you hear a search party's calls, shout only during the silence that follows.

IN THE EVENT OF HAVING TO ATTRACT THE ATTENTION OF A RESCUE PARTY, BUILDING A CAIRN IS A HIGHLY EFFECTIVE METHOD OF MARKING YOUR LOCATION OR DIRECTION OF TRAVEL. ENSURE THAT IT IS NOT MASKED OR COVERED OVER AS A RESULT OF BAD WEATHER CONDITIONS.

High-tech rescue options

Radio communication

Most small boats and all aeroplanes are fitted with two-way radios. If possible, send a 'Mayday' (French for *m'aidez* – help me) or SOS message before you crash or are stranded. When the radio set is on and tuned properly, transmission usually takes place by pushing a button on the hand-held microphone. In case of distortion, a weak signal or low battery power prevent you from transmitting a clear voice message, use the button to send an SOS in Morse code: short-short-short-long-long-long-short-short-short (see also p42) as 'short burst' signals can be picked up more easily than voice. On VHF (very high frequency) radios, Channel 16 is usually used for distress channel messages.

VHF transmitters may only be able to transmit signals within line of sight. Wait until you can see a search aeroplane or boat before wasting power, or move to a higher point.

Modern boats are often fitted with a rescue transponder that sends out signals. Many transponders are linked to SatNav (Satellite Navigation) systems that indicate the precise position. Check it is switched on.

Cellular phone communication

If you only have a faint or intermittent cellular signal, try dialling the emergency number as this works at a higher intensity than normal calls, or send a brief distress SMS (short messaging service). If neither brings response, try turning your phone on and off in an SOS pattern. You might be heard by one of the military or civilian 'ears in the sky', which monitor all signals. Short Morse code-like signals are the easiest to separate from other background electronic 'noise'.

It might take a day or two before your distress message is analyzed, but it could be your pathway to rescue.

The Japanese rescue service urges lost parties to keep cellular phones on or to turn them on as soon as they hear aircraft, as all cellular phones regularly emit unique 'pings', or identification patterns. These help the cellular network locate you during the sending and receiving of calls on your number. The Japanese use planes with mobile detectors which 'ping' the phone number of the lost person(s) if he or she is out of normal receiver range. Hopefully other countries will soon follow suit.

Communicating your position

S O S 1 5 7 6 2 8

••• = = = ••• | • = = = | ••••• | = = ••• | = •••• | •• = = | = = = ••

If you know where you are on a map, or precisely where you are heading to and want to communicate your exact position to searchers, use a set of coordinates with your SOS message. A system of six-figure coordinates is usually used.

Example: SOS 157628 (• indicates dot, short = indicates dash, long | indicates pause). This system is invaluable if you are using a lamp to signal an aircraft while you are on the move, or if Morse code is your only means of communication.

Disrupting telephone cables

This is definitely a last-resort method, but one that has brought swift results to people trapped in blizzards in wilderness areas that are traversed by phone lines.

Tapping the telephone lines: If you have a speaker of some sort, connect its two leads to bared sections of the two wires of a telephone line. Tapping a speaker cone can send impulses down the line, which should cause a phone to ring. Once answered, you may be able to listen and talk via the speaker, or tap out SOS signals. Phone lines do not operate at dangerous voltages. Note that this method will not work on modern fibre optic lines.

Cutting the line: This will bring the repairmen out! Most phone companies have technology that allows them to calculate where the break in the line is. Do bear in mind that this is a serious move, and will inconvenience or even endanger many people.

THE HEAT TRACE GIVEN OFF BY A LARGE, TIGHT-KNIT GROUP IS MORE EASILY DETECTABLE THAN THAT EMITTED BY AN INDIVIDUAL.

Battery tips

- Preserve all possible battery power for emergencies or for signalling when you see or hear a search vehicle approaching. Turn off or disconnect sets during long waiting periods.
- Most devices need a specific voltage to operate; if you want to increase the voltage, link batteries in series (positive of one to negative of the next, e.g. four 1.5V penlight batteries give 6V), and in parallel (all pluses linked to one another or all minuses) to increase power but keep the same voltage.
- Many devices such as electronic games, electric razors, pagers and laptop computers have batteries. If you are reliant on a small short-wave transmitter or a cellphone to transmit your distress signal, you may be able to power it by combining batteries from other devices. Make sure the voltage is right.
- Small batteries will be drained quickly if used to power a larger transmitter; however, a parallel circuit of several batteries might be sufficient.
- Keep batteries warm – place them in a sleeping bag or close to your body. Never heat them on a fire. Beware of substances leaking from batteries.
- Limit distress calls by keeping them short, and establish regular call times (i.e. every two hours); if searchers are only picking up your call signals at the limit of their frequency range, this will enable them to anticipate your call in order to trace your location more accurately.
- Solar battery chargers charge cells in a few hours in the sun. Many yachts have these, as do some caravans and aircraft – and even some children's toys.

Infrared traces

Huddling together in freezing conditions does more than share warmth and reduce individual heat loss; it also creates a larger infrared (heat) trace. By using infrared radar devices and goggles, civilian and military sources are able to detect people on the ground.

Being rescued

Helicopters are often used for final rescue operations. They do, however, have limitations. They require an obstruction-free entrance and exit path and a clear landing spot, and cannot touch down if a slope is too steep as their blades will hit the side. Even turbojet helicopters have minimal control at high altitude. The diagram below illustrates the preferred conditions for a helicopter to touch down safely. An improvised windsock has been created from a pair of long pants, indicating wind direction, and the helicopter is preparing to land facing into the wind. The aircraft is directed, by a large 'H' imprinted into the ground, to a clearing that is relatively flat and obstacle-free. The persons being rescued are safely sheltered away from the rotor blades and the downdraft they produce, while their bags have been anchored with rocks.

FLASHING MIRRORS AND BRIGHTLY COLOURED CLOTHING WILL ENABLE THE HELICOPTER CREW TO LOCATE YOUR PARTY.

ASSIST THE PILOT BY DEMARCATING A SUITABLE, CLEARED TOUCHDOWN POINT WITH A LARGE, VISIBLE H-SIGN.

Landing zone requirements

Pilots prefer not to lower vertically into a clearing, but aim instead for a low-level horizontal approach, usually into the wind. A suitable landing zone requires:

- A flat clearing with less than 10-degree slope and a diameter of 30m (100ft).
- No high obstructions (trees, large rocks) that may prevent an angled take-off.
- No telephone or power cables nearby.
- Clear away any twigs, branches, gravel. If snow-covered, try to compact snow in landing area.
- Indicate wind direction (makeshift windsock, smoke indicator, hand signal i.e. stand with back to wind, both arms pointing forward).
- Signal to the helicopter by flashing mirrors, wearing colourful clothing, or etching a large H into the ground to mark the touchdown area.

Precautions around helicopters

- Approach the aircraft from the front or side only AFTER you have been clearly signalled to do so.
- On engine shut-down, wait until the blades have totally stopped spinning – drooping rotor blades make a neat job of decapitating the unwary!
- Never approach or leave a helicopter on the uphill side; blades can spin very close to the ground.
- As you move towards the helicopter, crouch down and do not have any loose articles on you (i.e. hats, sleeping bags or ropes).
- Beware of holding long items vertically (e.g. folded stretchers).
- If it is impractical or difficult for the helicopter to land, the pilot may use a cable winch to lower rescue personnel or lift a patient – the winch sling can carry a massive electrical (static) charge from the aircraft, so always allow the winch to ground or touch water before grabbing it.
- NEVER fasten the winch to a solid object (i.e. a yacht or tree) or even to a stretcher, unless you have been requested to do so, because the helicopter may have to break away at any stage.
- If you are using the lifting strop, raise your arms, slip it over them, fasten the grommet (if it has one) and tuck it under your armpits.
- Give a thumbs-up signal before you link your hands. It is important NOT to raise your arms again after giving this signal.

DUE TO THE PROBLEMATIC ANGLE OF THIS SLOPE, AN OPTION FOR THE HELICOPTER IS TO DROP A WINCH TO THE PATIENT TO LIFT HIM OUT.

ROUGH SEAS ACCOMPANIED BY HIGH WINDS MAKE SEA RESCUES VERY TRICKY; IN THESE FREEZING CONDITIONS, THE ONSET OF HYPOTHERMIA IS RAPID.

Sea rescue craft

Rough water with large waves or swells could make it difficult for a rescue craft to reach you. You may be signalled to jump into the water. This might seem daunting, but let yourself be guided by the expertise of the rescuers. Ensure that you are wearing a life jacket and that it is fastened before moving into the water. If there are several people being rescued, try to stay linked together.

If being picked up from a boat by another craft, stand at the rail on the lee side (away from the direction from which the wind is blowing) until the other craft draws alongside. Then quickly climb over your rail and try to grasp the rail or ladder of the other craft before releasing your grip.

Jumping the gap should be seen as a last resort. A group should not swamp the rescue boat – try to stay calm and approach the boat in an orderly fashion. It is vital that one person takes command to avoid panic and coordinate actions of the people being rescued.

Meeting hostile groups

Although being noticed is usually a key priority, you may need to avoid contact with a hostile group or individual. You may find yourself in the middle of a guerrilla war, receive unwanted attention from drug or arms smugglers, or be part of a kidnap for political ransom. If you encounter a group and doubt their intentions, observe them carefully before approaching them. You can learn a great deal by simply watching their behaviour patterns.

If you are still unsure, rather send only one or two members of your group as messengers and caution the rest to remain hidden. Messengers should be simply dressed, and all visible signs of wealth should be left behind. Should they be captured or detained, hopefully you will be able to summon help to free them at a later stage.

Survival Priorities

Key survival priorities are shelter, clothing, water and food. Discussed here is what to take when travelling in different wilderness areas, how to react to the initial shock of a disaster, how best to use whatever you have with you and how to improvise if you find that you don't have all the necessary basics.

Taking action

By far the biggest killer in extreme situations is panic. This leads to irrational and ill-considered actions, or equally hazardous inactivity. Avoid heading out immediately. First pause, and review your options. It would be pointless building an elaborate shelter next to an aeroplane wreck only to find that there is a danger of fire, and foolish to move without first ensuring that you have taken all essentials.

The order in which you try to deal with the main priorities will vary, but in general, personal and group safety should always be high on the list.

Injured or trapped people Temper heroism with reality – in vehicle accidents, be aware of fire risks or collapsing parts of the structure(s). Deal with this without further endangering yourself or the group.

The group Gather them together. Ensure that everyone understands the situation and feels involved in subsequent planning, and give distraught or anxious individuals reassurance. In the aftermath of an accident, many people may still be in a state of shock. If this is the case, organizing clothing and finding shelter become your first priorities.

opposite AN IGLOO WON'T EXACTLY OFFER WARMTH, BUT BY CUTTING OUT ICY WINDS, THE TEMPERATURE WITHIN REMAINS CONSTANT.

Salvage Obtain what you can safely retrieve from any wreckage, but don't put anyone in undue danger.

Basic needs Only once your group understands the situation and has clarity on individual roles, should you look after the basic needs of group members according to the circumstances.

Clothing

Although one may have packed clothing to accommodate worst-case scenarios, this ideal is seldom realized, and improvisation is key. Cold conditions – either at sea or on land – pose by far the greatest danger. However, extremes of heat and humidity can also create problems to those without appropriate clothing.

Seat covers with holes cut in them make good jackets; carpets provide insulation and warmth; thin foam mattresses can be wrapped around the body; and parachutes, sails and tents all can be used to keep off the sun, or insulate against cold and wind.

above IN AN AIR DISASTER, ONCE THE SAFETY OF ALL THE PASSENGERS HAS BEEN ASCERTAINED, EFFORTS SHOULD BE MADE TO SALVAGE AS MUCH USABLE MATERIAL AS POSSIBLE FROM THE WRECKED AIRCRAFT.

Clothing for cold climates

Succumbing to the cold is a killer, and wet and windy conditions intensify the effects of low temperature. Always take enough clothing to cope with unpredictable weather changes, particularly at sea and in high mountain areas.

Follow the principles of layering. Multiple clothing layers trap warm air, providing effective insulation against the cold. You can easily adjust your temperature levels by adding or removing layers.

Inner layer – a warm, absorbent layer of polypropylene or a similar 'wicking' fabric which carries moisture away from the skin. A T-shirt is a good second-best.

Middle layer(s) – usually a thick cotton, nylon or preferably fleece shirt and a good-quality down garment, which is very light relative to its insulating value. Other options are down, wool, polar fleece or similar thick-pile fibres.

Outer layer – wind- and waterproof; the best material is a breathable fabric, such as Gore-Tex®, which reduces the build-up of perspiration around your body.

Extremities – two pairs of socks (one thin and one thick), gloves and a snug balaclava. Up to 25 per cent of body heat can be lost via an uncovered head and neck area, and a further 20 per cent through the hands and feet.

Finding shelter

Sun, wind and cold all affect the body and you will need to find protection from these. If you are fortunate, you may have a tent or find natural shelters such as caves, overhangs or hollows nearby. Otherwise you need to improvise. Personal security is closely linked to having a roof over one's head; as survival focuses so much on one's state of mind, shelter should always be seen as more than merely a physical necessity.

Tents

A good tent enables you to camp under extreme conditions. By building windbreaks out of rocks, logs or banks of earth or snow and using these to help anchor the tent, it is possible to withstand high winds. Even heavy rain and snow should not be a real problem provided you dig channels alongside the tent to drain the site, and regularly clear away snow before it accumulates on the tent and its weight breaks the poles.

Ideally, you should pitch your tent away from a steep slope, but not in a basin, which could flood. If you are near a river, camp above any possible flood levels. Water can build up much further upstream and become a raging torrent as it progresses downstream towards you. Take note of wind patterns and try to shelter the tent by placing it in the lee of bushes, rocks or trees. When camping in snow, site your tent well away from any potential avalanche zone.

Basic tent types

a. Geodesic – good in mountains with high winds and snowfall; has the highest strength-to-weight ratio. Three- or four-pole versions are more stable than two.

b. A-frame – also known as a ridge tent, it has plenty of headroom and is often less expensive than a geodesic, but not as stable in wind.

c. Tunnel – fairly stable and lighter for its size than the geodesic, but not as good in heavy snow.

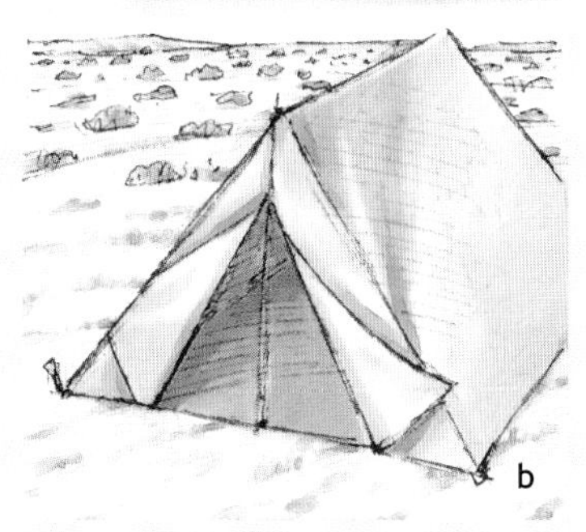

Natural features

Hollows A hollow or a space under a rock can provide welcome shelter from the wind. Roof it with small branches and any other suitable material.

Logs and fallen trees A lean-to shelter can be formed by scraping out a small depression on the lee side of a log and placing branches or twigs over it.

Under snow-laden trees Natural depressions form in the snow under the lower branches of trees. Enlarge them by digging down into the snow drifts.

Cave or deep overhang Check for tracks at the entrance to ensure that the cave does not contain an animal of some kind. Cover the entrance with brush, branches or other material to shield you from the elements. Ensure that you have adequate ventilation.

A ROCKY OVERHANG PROVIDES SHELTER FROM THE ELEMENTS AND ALSO ACTS AS A PARTIAL BARRIER TO ONSHORE BREEZES.

Ice constructions

Shelter from the wind is often a high priority. Inside a snow shelter the temperature can remain constant, unlike the plummeting temperatures in the open. Life-saving options include a trench dug in the snow and roofed by a groundsheet or branches packed with snow, and a snow cave dug in a bank or glacier.

Building a quinze

Simple igloo (see illustration) made of compacted snow:

- Form a pyramid about 1m (3ft) high from heaped backpacks or equipment. Pack snow onto this, compacting each 5–10cm (2–4in) layer well and allowing to freeze for 20 minutes.
- When the dome is around 1.5m (4ft) high and 2.5–3m (7–10ft) in diameter, push in short sticks (or similar) to a depth of 25cm (10in).
- Dig an entrance and remove the core of snow until packs/equipment can be removed. Hollow out the quinze until you reach the sticks. Compact the snow from the inside.
- Once inside, use a pack to block the entrance and preserve warmth.

Constructing a quince

In the desert

Dry air and lack of insulation result in extremely hot days and bitterly cold nights. Shelter from the sun is a priority – use a sail, groundsheet, piece of cloth or any other available material such as a sleeping bag. Lift the shelter off the ground by piling up sticks, stones or your backpack as a base to allow air flow. If you have no cover, make use of even the smallest amount of shade cast by rock outcrops or desert plants, no matter how dry and scanty. In terms of digging, the sand will be very hot, but if you dig fairly deep, you will reach cooler sand. Climb into the hole and cover yourself with a sand layer to shield your exposed skin from the ultraviolet (UV) rays.

This method can also be used at night to shield you from the cold, but first wrap yourself in a sheet or canvas if you have one available.

If you need to shelter from a desert sandstorm, sit with your back to the wind and cover your body. By wrapping a cloth loosely around your head, you will still be able to breathe while shielding yourself from the stinging sand.

A SCARF, SHAWL OR PIECE OF CLOTH WRAPPED AROUND THE FACE IN A SANDSTORM ALLOWS YOU TO BREATHE BUT FILTERS OUT THE SAND.

Tropical and jungle shelters

Most shelter construction will rely on improvisation, depending on the circumstances and materials available. You may need to modify the ideas given here.

Tepee

Lash six to eight longish poles together at one end (the top), raise the rough structure upright and spread the poles evenly apart. Digging small holes for the pole bases will stabilize the structure. Cover with canvas or other material; use small branches and overlapped leaves if nothing else is available.

A-frame construction

You can create an effective waterproof A-frame construction by using a canvas sheet, small branches and leaves, or grass. In warm regions, bamboo makes excellent building material, but be careful as it can create dangerously sharp splinters when it is cut or split.

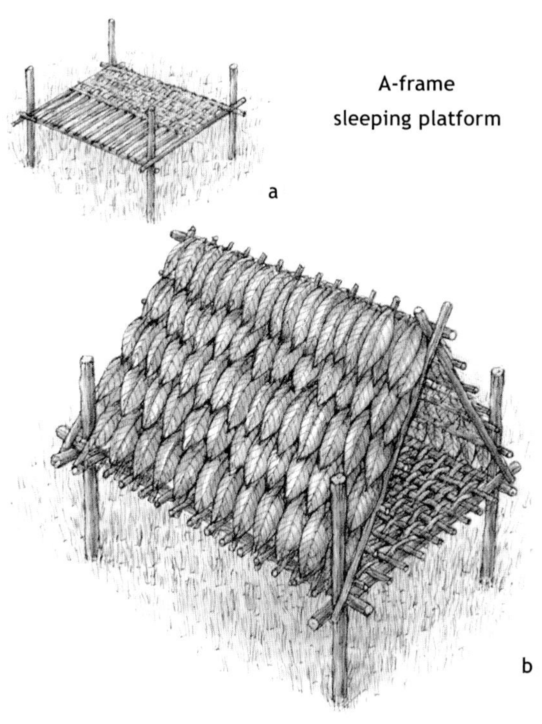

A-frame sleeping platform

DERIVED FROM THE CONE-SHAPED ANIMAL-SKIN TENTS USED BY NORTH AMERICAN INDIANS, THE TEPEE IS ONE OF THE SIMPLEST STRUCTURES AND STILL APPEARS TODAY IN THE FORM OF MODERN TENTS.

Sleeping platform

It is best to elevate your sleeping area above the ground, away from the damp and any crawling insects. Starting with four vertical poles, lash horizontal support poles across the ends and sides to create a basic frame for a sleeping platform. Then add numerous crosspieces (a) on which to place your makeshift mattress of leaves and branches or other suitable material. Take the basic A-frame (or even tepee), and attach it to the four corners of the platform (b).

Hammock

The net-making method on p65 can be used to make a hammock, which is particularly valuable in preventing bed (pressure) sores if you have a disabled or ill group member who may need to spend a few days lying down. Hammocks are handy, too, for keeping sleeping bags off damp ground and to avoid crawling creatures. Attaching spacer bars at each end (these can be pieces of wood or metal) will ensure that the hammock does not fold around you.

Water

Water is far more important to the body than food. We can go for quite a number of days, even weeks, without food, but for only a few days without water before our vital functions begins to deteriorate.

Under normal conditions, a moderately active person needs 2–4 litres (4–8pt) of water per day. If you find yourself in a situation where it appears that you may run out of water, never assume you will be rescued quickly – take an inventory of all water and other liquid supplies and enforce strict rationing.

Conserving body fluids

- Staying slightly 'thirsty' right from the start will diminish your urine output and aid water retention later, when your body needs it most.
- Even in very cold conditions, a good deal of water is lost while you are breathing because of the dryness of the air. Make a point of breathing through the nose rather than the mouth.
- Restrict your activity to an essential minimum to help preserve your body's water content.
- In hot conditions, try to perform strenuous activities at night. Stay in the shade in the daytime.
- Avoid alcohol and caffeine.
- Avoid dry, salty foods.

PITCHER PLANTS CARRY WATER BOTH IN THEIR FLOWERS AND STEMS.

Sources of water

Environment

Rainwater is almost universally safe, except after a large volcanic eruption or huge fire. Collect rainwater by any means possible and store as much as you can.

Water from ice and snow Ice is easier and quicker to melt than snow. If forced to use snow, then dig down as deeper layers are more granular and provide denser snow. Sea ice has a good deal of salt in it, except for older weathered ice (such as icebergs), recognizable by a blueish tinge. If there is no fuel to melt the snow, form it into compact balls. These should either be exposed to the sun or placed next to your body in some form of waterproof container, ideally a black plastic bag. Remove the ice ball from the plastic and suck water from the bottom of the ball.

Jungles and tropics

Most jungles and tropical areas will have rainfall, often on a daily basis – collect this. Foliage from large-leaved plants can be tied together so that the tips angle downwards into a container. Many larger-leaved jungle plants have water trapped where the leaf meets the stem, including some of the large and attractive orchids, as well as bromeliads and pitcher plants.

Palm and banana trees have plenty of water in their trunks. Cut a small banana tree off close to the ground and make a hollow bowl in the stump with a sharp object. The hollow will slowly fill with safe clean water. The water from palm tree stems is most easily accessed by cutting the tip off a flowering stalk and bending it downwards. Coconut milk is safe, but in ripe coconuts can have a strong laxative effect.

Water can be obtained from bamboo stems by cutting a notch at the bottom of a section and letting the water drain out. Older bamboo often holds more free water than young, green bamboo.

Some vines (the round, not flat types) have drinkable water in them. Avoid those with a milky sap, as this is usually poisonous. Cut a notch as high as possible to release the water, then cut the vine low down to allow the water to drain out.

Finding water in dry areas

Search at the lowest possible point. Be prepared to dig deep in seemingly dry riverbeds. You will have a better chance digging where the riverbed turns at an outer bank, or meets a rocky outcrop.

On beaches, dig above the high-tide mark. Give water time to trickle in – it may be brackish, but is certainly drinkable.

Observing animals and birds

Follow animal tracks or the flight pattern of birds (this is tighter and more organized when flying to water). Most animals and birds drink in the early morning hours or late in the afternoon. If tracks lead downhill and converge, they could lead to water.

Larger fish often have water alongside their vertebral column; to retrieve it carefully gut the fish and remove the backbone.

Arid region plants

- Cacti and aloe species are excellent water sources.
- Plant roots often contain stores of water. Dig down and cut the root, then crush it; extract its water by squeezing the pulp through a cloth.
- The Australian Aborigines collect moisture by tying bunches of grass to their ankles, then walking through the dewy grass fields in the early morning.

STUDYING ANIMAL PATTERNS MAY LEAD YOU TO WATER SOURCES.

Water purification

The concept of a 'pristine mountain stream' is almost a modern myth, as even some of the world's most isolated streams carry considerable debris resulting from careless human ablution. Human faeces harbour micro-organisms such as Giardia as well as other problem bacteria and viruses. Take note of the following tips:

- Obtain your water from a point as close to its source (spring or clean snowmelt) as possible.
- Most portable water filters do not remove all bacteria and micro-organisms, in particular viruses. Ensure you buy one that is 100 per cent effective.
- Boiling after filtering is effective in most cases, but not totally so, particularly at high altitudes. If fuel allows, boil for 10 minutes at sea level, adding 1 minute for every 1000ft (300m) altitude gain.
- Filtered and boiled water should then be further sterilized by using purification tablets to make it completely safe (either chlorine or iodine water purification tablets).
- Another emergency purification alternative is to add five drops of household bleach to a litre of water, and allow to stand for 45 minutes. It may have a strange taste, but should be safe to drink.
- Leave water in a shallow container in bright sun for a few hours to kill many bacteria and viruses.

Filtering water

The first step in water purification is to filter out debris by using a good commercial filter, or devise your own from coffee filter papers, car air filters, handkerchiefs or items of clothing.

If all you find is mud instead of water, place it in a piece of cloth, and wring the moisture out into a container. Any water-containing vegetation can also be strained through a cloth.

You could dig a water hole in the sand; prevent it from collapsing on itself by lining it with plant material. This has the added advantage of preventing small animals from making use of the water hole.

WATER SQUEEZED FROM MUD OR VEGETATION IS FILTERED THROUGH THE CLOTH, BUT SHOULD BE PURIFIED FURTHER.

Stills to purify water

A good still can be used to make seawater, other questionable water and liquids such as radiator fluid safer to drink, however unless the still is 100 per cent efficient (highly unlikely in survival conditions) the water will not be totally pure, and may still have a strange taste and/or harbour germs.

If you have no sophisticated means of purification, no matter how bad the water looks, filter as well as you can using basic stone and sand filters, before drinking. Any diseases can, in all likelihood, be cured after you have been rescued, and at least you won't die from dehydration!

Water from transpiration

Most trees and plants transpirate ('sweat') water daily as part of their natural water and food transport system. To take advantage of this handy water source, tie a large plastic bag around a leafy branch or section of a leafy bush, without tearing the bag and sealing all openings well. Make sure one corner of the bag hangs low so the water can collect here.

DEW THAT COLLECTS ON LEAVES IS AN EXCELLENT SOURCE OF WATER, AND ONE USED BY MANY WILD CREATURES.

Passive (solar) still

- This method collects water from plant transpiration.
- Dig a hole roughly 40cm (15in) deep and 50cm (20in) wide; fill it with leaves and other leafy vegetation.
- Place a small container inside the hole; this acts as a receptacle for the fluid to be collected.
- A piece of plastic, weighted with stones, dips towards the receptacle; liquid evaporates and condenses onto the plastic, then drips into the receptacle.
- Alternatively, to condense and purify seawater, a small container can be placed in a larger one holding the seawater, to act as the receptacle for the distilled fluid.

Active (steam) still

In the event of a vehicle wreck or breakdown, you might well have the materials to produce a workable steam still. Any cracks or joints in the distillation tube can be sealed using mud or clay.

- A semi-sealed container contains liquid to be distilled and is suspended from a sturdy tripod over the fire.
- A distillation tube runs from container to distillate bottle.
- Cooling (sea) water (top container) speeds up condensation by dripping onto a jacket wrapped around the distillation tube to distil the liquid more efficiently.

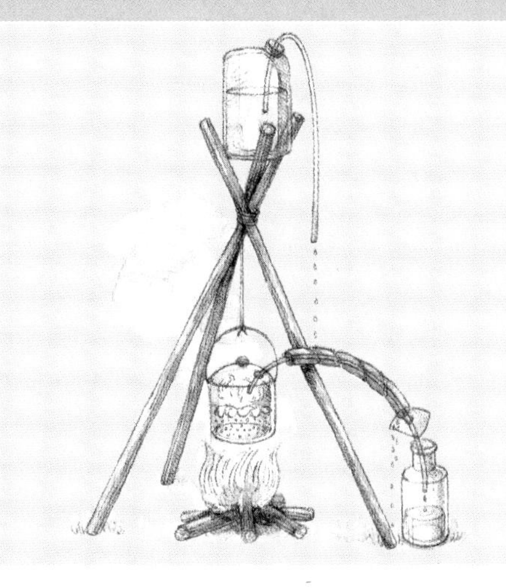

Plant still

- You need a small bush or you can gather a pile of freshly cut vegetation to create a 'bag still'.
- Select a large plastic bag, if possible, and tie it tightly around the small bush, or around a branch with a good supply of leaves.
- In you use cut vegetation, place twigs so that they keep the bag from collapsing; for the small bush, surround it with the bag, ensuring that all openings are properly sealed.
- Water will collect in the low-lying corners of the bag.

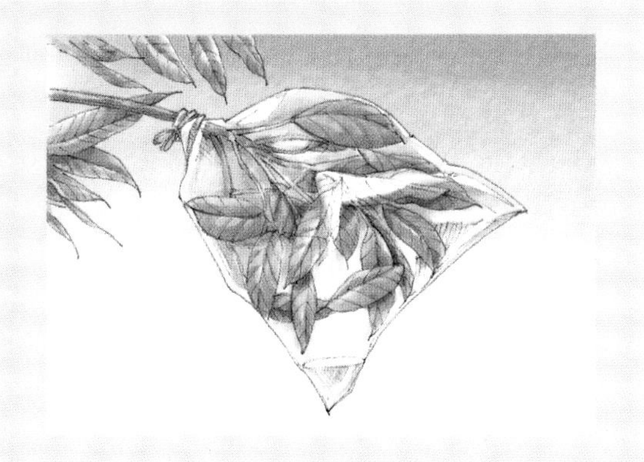

Obtaining food

Your body is capable of going without food for long periods of time (records exceed 75 days). In a survival or potential survival situation, food should be rationed very carefully by applying the 'worst case' principle. Some groups experience tough times due to unrealistic expectations about being rescued.

Testing plants

- Allocate only one person in a group to test each plant type.
- Test each part of the plant (roots, stems, leaves, etc.) independently.
- **Smell:** if a crushed plant smells of almonds (hydrocyanic acid) or peaches (prussic acid), discard it.
- **Skin reaction:** rub a piece of the crushed plant on soft skin (inside of arm), and wait five minutes. Check for rash or burning.
- **Mouth:** place a small piece on the lips, then the tip of the tongue. Chew a small amount without swallowing. Watch for numbness, stinging or burning.
- **Swallowing:** chew and swallow a small amount, then wait for at least three hours before eating anything else. Do the same tests after cooking plants – some chemical substances may have been altered.

NUTS ARE ONE OF THE SAFEST SOURCES OF FOOD IN THE WILD.

Locating food from plants

Palm trees – coconuts and other palm fruits as well as the soft heart of the young stem or branches can all be eaten.

Pine trees – when dug out of the cones, the seeds of this tree form a tasty, nutty food source. The leaves can be boiled to provide a type of tea.

Lichens – this safe, nonpoisonous food needs to be boiled to soften it. Lichens can also be used to create a watery soup.

Sea lettuce – there is no truly poisonous seaweed, although some can cause stomach irritation. The common sea lettuce and kelp found along beaches around the world make safe and nutritious meals. They should first be washed, then boiled. The leftover water makes a nutritious soup.

Nuts and nut-like seeds – these are an excellent source of much needed protein. Very few nuts are poisonous or harmful.

Fern species – many types of ferns, particularly those found in the Northern Hemisphere, can be eaten. None are poisonous except for mature plants of the most common species, bracken fern (*Pteridium aquilinum*). Eat only young ferns – those with tightly coiled fronds. Before you eat them, remove the irritating little hairs that cover many fern species.

Animals as a food source

Mankind can survive quite happily on a completely vegetarian diet, and it is often the easiest food to obtain. It is worth remembering that any plant a monkey (but note, NOT a baboon) eats is usually edible to man. This is not true for other animals and birds; many consume plants or seeds that can be harmful to humans.

Otherwise, you may need to take the decision to turn to animals. Smaller animals (including insects, grubs, marine invertebrates, molluscs and reptiles, as well as fish and birds) can be easier to accept as a food source and are often simpler to trap or catch than the larger animals. Grubs, which tunnel inside dead or decaying trees, taste like roasted peanuts when fried.

Make a point of avoiding very brightly coloured insects, as some are poisonous.

TO MAKE INVERTEBRATES MORE PALATABLE, IT IS ADVISABLE TO REMOVE THEIR WINGS, LEGS AND HARD OUTER CASING.

Animal patterns

Animals are creatures of habit, following set patterns of feeding and drinking, and can often be traced to permanent burrows or homes. Examine paths through the bush – however faint – to establish whether animals regularly use them, and obtain information on the size, numbers and even the type of animal that uses them. It would be far easier to catch an animal on its regular route or trap it in its nest or burrow than to stalk it in the open. Hibernating animals in colder areas are fairly easy to remove or dig out once you have found their lair.

Small reptiles and amphibians – with the exception of toads – are all edible. Some have glands in the skin that secrete a harmful mucus, so remove the skin of these animals. Lizards, geckoes and snakes are excellent potential food sources. In the case of snakes, cut the head off well behind the mouth to avoid contact with any poison glands.

IN SOME FISH SPECIES, MUCUS ON THE SKIN MAY BE TOXIC; IF IT FEELS SLIMY WASH IT THOROUGHLY, OR RUB WITH SAND BEFORE COOKING.

Snares and traps

The aim of this section is not to encourage people in non-survival situations to head for the wild, build traps and snares, and catch an animal. Prior knowledge of how to construct these traps, however, would be very helpful in a real survival situation. If you do practise your skill and accidentally catch an animal, treat it with dignity. Try to avoid injury, and release it if you can. Dismantle any snare or trap before you leave the area.

Snares and traps can be made with various materials and devising your own version is a key element. Wire, flex, cord or string can be used to build many different types. If you do not have any, you may need to make deadfall traps, pits or balance traps. Correct balance and tension of deadfall traps requires much patience. Bait the trap with your prey's favourite food type if you can.

Weigh up food gain against the energy required to trap and hunt. You may decide to abandon these options, rather. Take great care when setting traps. Triggers need to be responsive, but a number of hunters have had their creations fall on them instead of the intended animal!

Simple loop snare

- Twist a running loop of wire or smooth rigid nylon into a small eye (or via overhand knot)
- Feed free end through the loop and tie to an anchor.
- The loop, whose diameter should match the size of the prey, must close easily.
- Use twigs to hold the loop open.

Ground-based spring snare

- A simple noose is attached to a strong branch/tree.
- A trigger bar is fitted to a short, straight stake.
- The loop is firmly attached to the trigger bar, which is attached under tension to a pliable bent branch.
- When triggered, the noose will whip up and tighten.

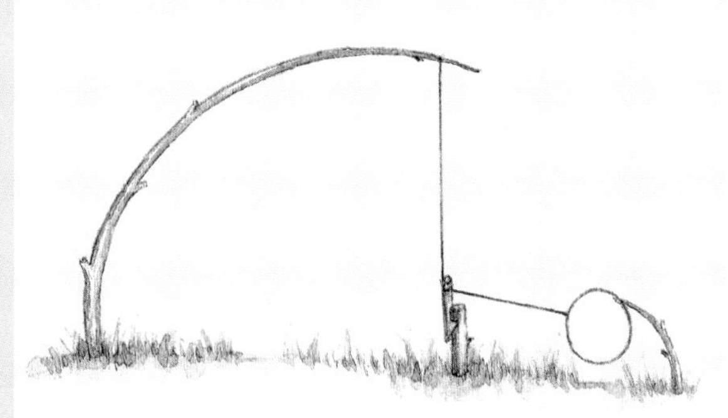

Deadfall trap

- Cord or string is attached to a baited trigger bar.
- When the animal pulls on the bait (or walks into the tensioned cord), it releases the deadfall.

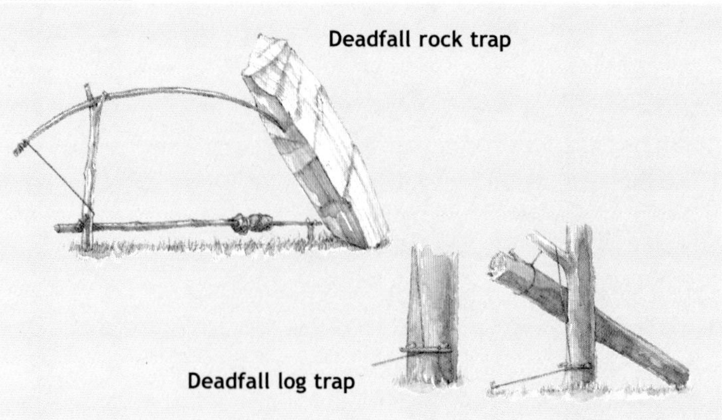

Hunting weapons

Catapults

Create a simple catapult from elasticized material such as a strip of rubber or clothing elastic. One Himalayan survivor rescued in a dire state of starvation admitted that, had he only thought of it, he could easily have obtained food from birds in the area. His pack contained all the materials to build a good, strong catapult.

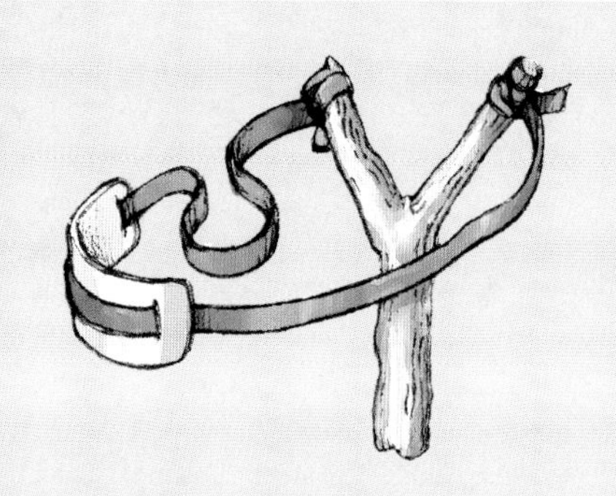

Slingshot

Easier to make than a catapult, this weapon needs practice. Using smooth stones, twirl the sling above your head at speed and release one thong in the intended direction. Make sure you have no one near you, as your aim will initially be unpredictable!

Spear

A spear can bring down larger prey such as a deer or buck. Fastening a knife or sharpened piece of metal to one end makes it more efficient. Practise throwing from a crouching position without having to stand up, as this alerts the animal.

Bow and arrow

A bow and arrow is probably the most useful weapon as it can kill at a great distance. It could take a lot of effort to make this weapon but the rewards are considerable.

Making bows

1. Select a hard, springy branch (e.g. young spruce, cedar or eucalyptus) and notch the ends.
2. Loop tough string around each end and tighten for tension.
3. Securely lash ends with string.
4. Fit an arrow of suitable length.

Making arrows

Sharpen a straight piece of wood (60cm, 24in) and notch the other end. A length of tent pole or even a section of fuel or brake line are also suitable; notch one end and try to sharpen the other, or attach a point. Vanes can be created from feathers, firm plastic, tough leaves or paper and tied in place for improved arrow flight.

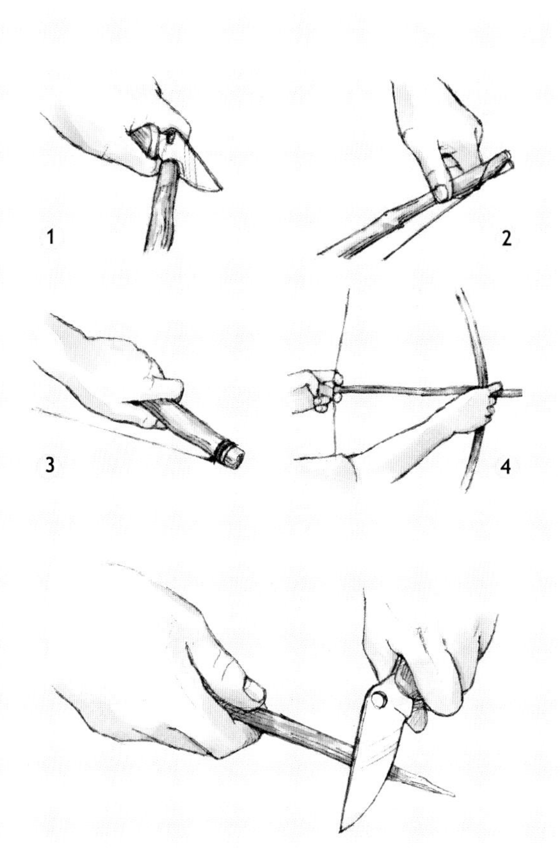

Fishing

Fish are a valuable food source, with a high protein content and large amounts of fat as well as vitamins and minerals. All freshwater fish and most marine fish are edible, although some are tastier than others.

Marine fish are attracted by any spinners that turn and flash in the water – try to create these. Numerous hooks can be placed on one long line, which can be dropped astern. Many fish are also attracted to the shade of the boat and will swim around it for hours.

It is often best to de-scale and de-bone fish before cooking and eating them although this is not necessary with smaller fish (e.g. sardines and minnows). If the skin of the fish feels slimy, wash the mucus off thoroughly, then rub the skin with sand to remove all traces and wash once more.

Certain common marine catfish have poison sacs attached to large spines on their fins, as do tropical species such as trigger fish, stonefish, scorpion fish and zebra fish. Some reef fish, including the porcupine and puffer fish, build up dangerous toxins especially but not exclusively in their livers. Generally, avoid reef fish that have 'parrot-like' beaks or slimy, mucus-covered skins.

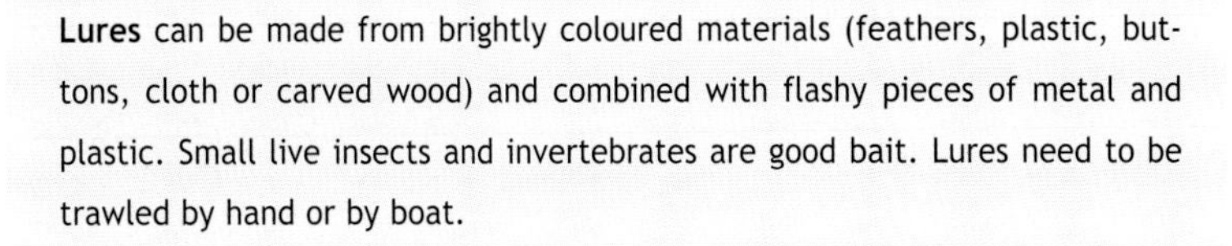

Lures can be made from brightly coloured materials (feathers, plastic, buttons, cloth or carved wood) and combined with flashy pieces of metal and plastic. Small live insects and invertebrates are good bait. Lures need to be trawled by hand or by boat.

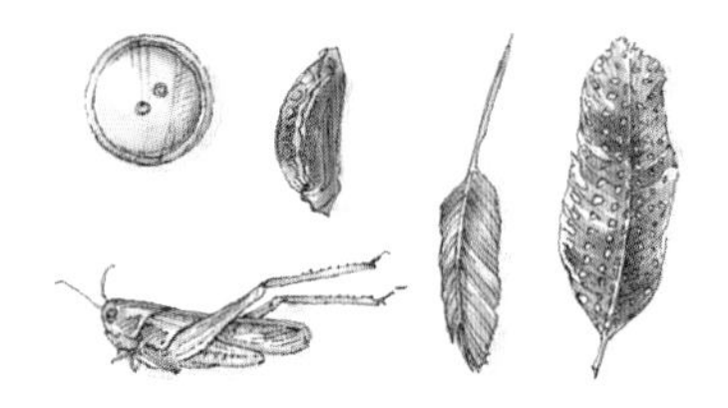

Improvised hooks can be made from a key ring, wire, safety pins, branches and thorns. Attach via a notch in the wood or a closed loop in the metal. File a point or sharpen on stone.

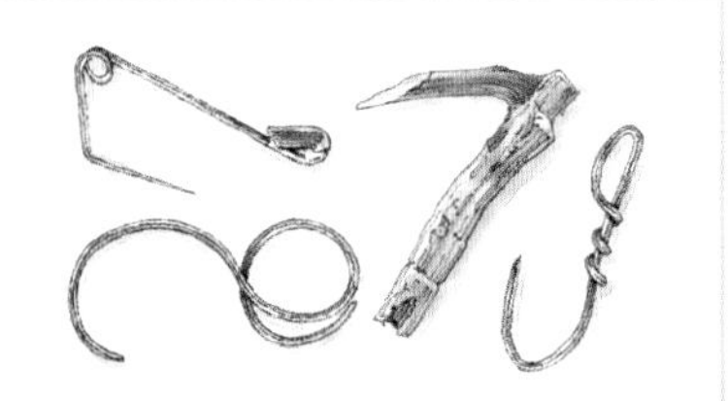

Useful fishing equipment

Fish traps

Traps are useful as they can be set and left, for later inspection.

Bottle trap

Cut the neck off a plastic bottle, invert and jam it back into the bottle. Bait can be placed in the main section. Small fish such as minnows are trapped once they have swum through the neck.

Twig trap (lobster trap)

A funnel trap has two parts: a small inner funnel that fits into a larger outer section, constructed of sticks tied together and made escape-proof with twine mesh. The smaller funnel ends in a narrow open cone whose sharpened spikes prevent fish returning through the constriction after they have swum in.

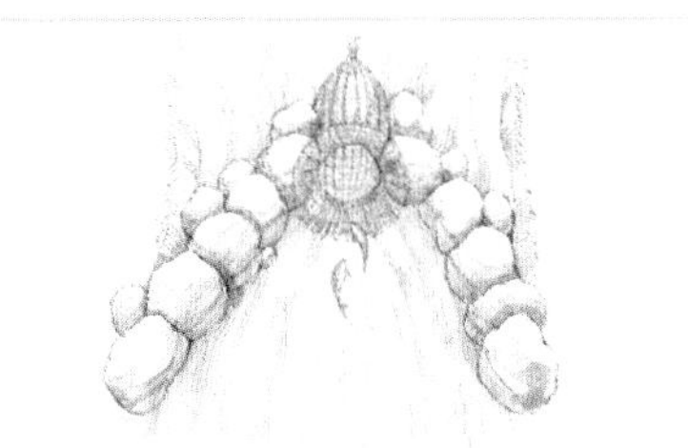

Sock trap

This trap made of a sock (or a pillow slip) is held open by a ring made of wire or the neck of a plastic bottle. Smelly bait (even animal entrails or dung) inside the sock will attract eels.

Using harpoons or running lines

To strike fast with a harpoon, start with barbs submersed in the water. Because water refracts light, aim lower than the spot where you see the fish. You are more likely to catch a fish by suspending a series of baited hooks on lines of different lengths, either from a bank or from a makeshift boat.

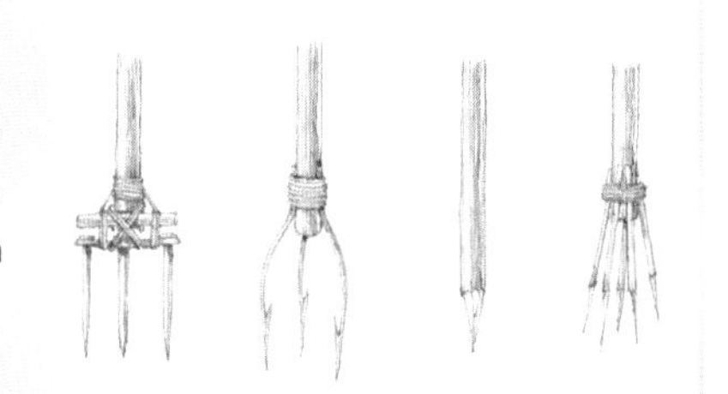

Ice hole with fishing sticks

In very cold areas where the water surface has frozen over, cut an ice hole making sure you remain on stable ice. If you can, set up several lines of varying lengths and bait.

Constructing a net

Nets have many uses – catching fish, collecting food and making hammocks – and can be made using any twine or cord. Cord should ideally not be too thin; parachute cord is a good option or even insulated electric wire. The easiest net to make is a tie-off net (see diagram).

- Strong top rope (or doubled or plaited thin cord) is suspended between two posts.
- Separate drop-threads of cord are tied to the top rope using cow (girth) hitches.
- Tie together adjacent drop-threads, sequentially, using overhand knots to create the tie-off net.
- The net is finished by tying double drop-threads to the bottom strong rope using a clove hitch (see p71) or round turn and two half-hitches.

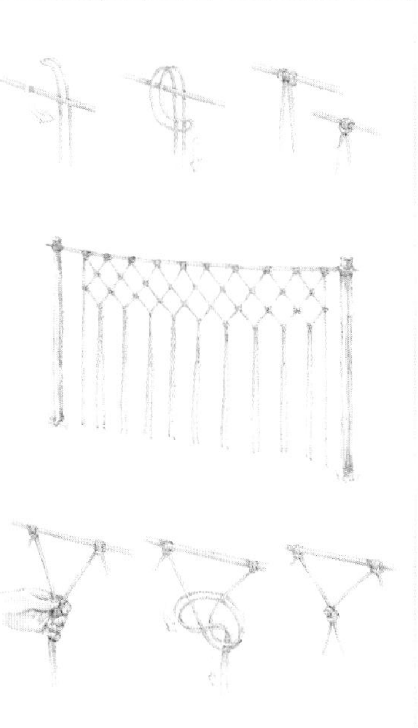

Fires and firemaking

A fire can make a tremendous difference to a group's physical and mental well-being – the difference between life and death. Making a fire is simple if you have a lighter or matches – waterproof matches are still the best. If you have a firemaking means, preserve it; once you have used it up, lighting a fire will become very difficult if not entirely impossible. Always take at least two firemaking methods into the wilderness!

Be very careful where and how you make a fire, as it can easily get out of control.

Firemaking materials

Tinder – Any dry, fine, combustible material such as paper, moss, tree bark, dry leaves, grass, animal dung or fungi will allow a spark to take. Crush or grind the material first to make it fine enough to flare easily. Shredded nylon or polyester from clothing makes good emergency tinder.

Kindling – Small leaves, sticks and pieces of dry bark are then added to the fire, done slowly and carefully until a clear flame is visible.

Main fuel – Avoid smothering the fire – add fuel progressively. Small sticks need to go on first, then large branches or logs as the fire takes. Greenish wood or wet fuel can be dried out next to a fire.

LIGHTING A FIRE WITH MINIMAL MATERIALS IN DIFFICULT CONDITIONS SUCH AS HIGH WIND, COLD OR RAIN IS EXTREMELY TESTING. WHEN YOUR SITUATION IS DESPERATE, PRESERVE THE TINIEST SPARK AND DO NOT SMOTHER A TENTATIVE FLICKER BY OVER-HASTILY ADDING FUEL.

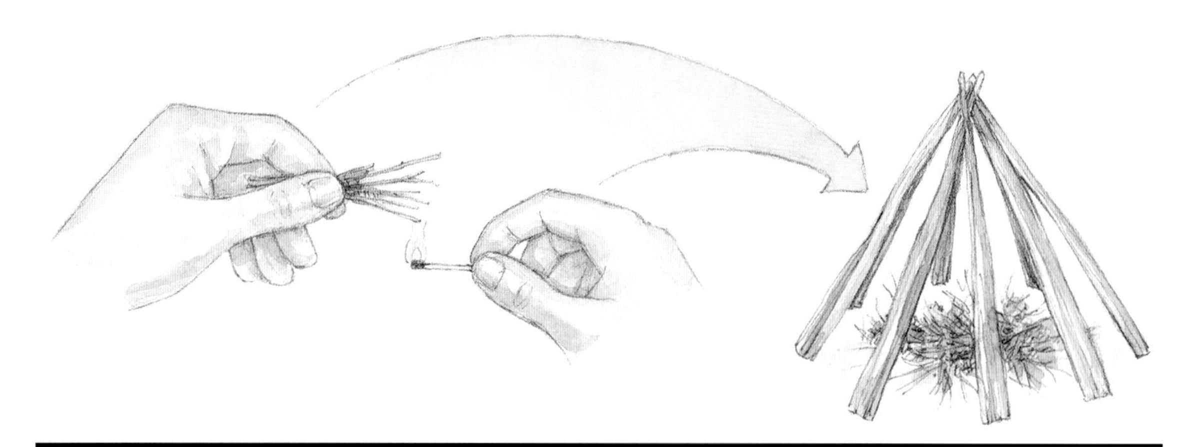

LIGHT A BUNDLE OF DRY TINDER BENEATH A PYRAMID OF BRANCHES, WHICH EVENTUALLY BURN FROM THE BUILD-UP OF HEAT.

Alternative fuels

Many substances will burn, particularly fuels, plastic, rubber and cloth. Some, especially plastics, can give off dangerous gases. Use them only as a last resort and ensure the fire is lit in a well-ventilated area.

Oil and petrol can be persuaded to burn more slowly by pouring these into a container of sand and lighting them. Oils and diesel fuel burn more readily when mixed with petrol. Hydraulic fluid and pure anti-freeze are also flammable. You can also get a 'lazy' fire to start by spraying from an aerosol can – aerosol propellants and contents burn well.

Lighting a fire without matches or lighters

The following methods are worth trying, but not easy!

- Strong sunlight, focused through a magnifying glass or fairly strong glasses, can start a fire. Binoculars and cameras can provide useful lenses if broken open.
- Flintstones generate sparks when they are struck against one another, or using a hard piece of steel. The sparks given off should be directed into a bed of very fine, dry tinder.
- Magnesium blocks are highly inflammable and are found at outdoor stores. The scrapings from these blocks provide superb tinder.
- A car or other large battery (even lamp batteries) can produce sparks to light a fire, particularly from steel wool or bunches of thin bared wire.

A BLADE STRUCK AGAINST A ROCK CREATES SPARKS THAT CAN BE DIRECTED ONTO BONE-DRY TINDER TO START A FIRE.

Different styles of fire

Apart from the normal pyramid fire, you can build fires that make more efficient use of available fuel, and are less hazardous. NOTE: Rocks may split or crack in a fire, particularly if they are wet or porous. Beware of flying rock pieces as they can cause serious injury.

Star fires – these have a circular arrangement of logs meeting in a central fire; the logs are pushed inward as they burn down. Rocks can be placed in the spaces between the logs to provide cooking platforms and to hold the logs in place.

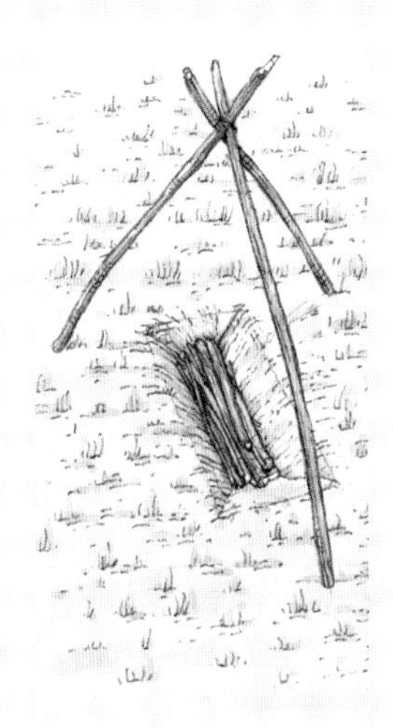

A trench or pit fire – placed in a sheltering hole, this type of fire is good in strong wind. It also reduces the danger of flying sparks. To aid ventilation, place a layer of fairly large rocks on the bottom and build the fire on top of them. One or more reflector shields will help you to direct warmth to you or the group.

A tin or 'hobo' fire – holes are punched into the base and around the lower half of a large tin, and a panel is cut near the base, allowing for a fold-back flap. This allows fuel to be inserted and to burn efficiently.

Holes can be punched into the top or a small disc cut out to accommodate a pot. The tin is placed on a stable circle of rocks.

Food preparation

Most food needs some preparation to make it more palatable or to remove poisonous or harmful sections. In a true survival situation, you would eat parts of animals that would usually be thrown away. In fact, very little cannot be eaten.

In many cases, food does not actually need cooking and the process actually destroys many vital nutrients such as vitamin C. However, cooking can make certain foods more palatable by softening meat and vegetable fibres, and will destroy harmful bacteria and parasites. Poisonous substances in some plants (such as old potatoes, nettles and leeks) are also neutralized.

Circumstances may force you to eat your food raw – you should do this with the knowledge that the human body is actually biologically adapted to a diet of primarily raw foodstuffs.

Preserving food

To prevent food from spoiling, it must be properly dried or preserved in pickling liquid.

- Preserve meat and fish by boiling, smoking or drying. Make a 'smokehouse' by closing off the sides of a tripod standing over an open fire.
- Cut meat into very thin strips, remove fat, then string or spear it onto sticks or racks.
- Salt rubbed into meat aids preservation (but could bring on thirst where water is scarce).
- The citric acid of lemons and limes is good for pickling fish, meat and vegetables. Use equal parts of juice and water.
- A strong salt solution can also be used as a pickling preservative; if a raw potato, tuber, or onion floats in the solution, the brine is strong enough.
- Smoke or sun-dry thin slices of fruit or vegetable.
- Lichens and seaweed need to be boiled first before smoking or drying. Once completely dry, these can be ground to a powder to be used as seasoning.
- Sun-drying can only be done in hot, dry climates.

Hot meals

If you are in a position to cook food and you have no utensils, cook directly on hot rocks. These can either be heated up in a fire and removed when needed, or the fire can be made on a bed of rocks. When the fire has burnt out, the rocks can be swept clean, and the food cooked on top of them. You can also wrap food in a suitably large, non-poisonous leaf (for an edibility test, see p60). Then plaster the leaf with mud and place this 'mud sandwich' on a bed of coals, layering more coals on top if possible. This method is a slow but effective way of cooking most meals.

ROCKS RETAIN HEAT WELL AND CAN BE USED TO SLOW-COOK FOOD.

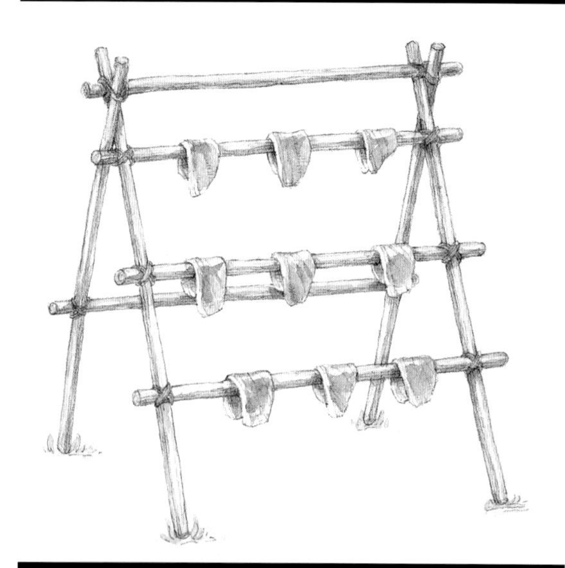

FISH OR MEAT STRIPS CAN BE DRIED ON THIS IMPROVISED DRYING RACK.

Camping stoves

Standard camping gas stoves – Best for lightweight camping trips, they are compact and easy to use. They do not burn well at extreme altitudes on normal fuel, but have been used with special propane-butane mixes up Mount Everest. They light easily, with a controllable flame. Cartridges must be empty before they are removed (never throw them into a fire). Do not change cylinders in the vicinity of open flames.

Gas stoves with removable cartridges – They can be stored and carried separately from the stove, and can be used on a matching gas lamp.

Meths stoves – These use methylated spirits or ethanol (known as 'cooking alcohol' in many parts of the world). Most come as compact fold-up units, complete with kettle, pots and windscreen.

Multifuel stoves – Fuelled by petrol, benzine, paraffin, and even alcohol, they tend to be temperamental in terms of flare-ups or blocked jets, but are nonetheless very efficient. It is wise not to use these stoves in tents or other confined spaces as the pressurizing system is not always completely reliable, especially after periods of heavy use.

In many areas (such as large parts of Africa, South America and Eastern Europe), mini gas cylinders or specialized fuels may not be available.

Fuels are poisonous, and all are highly inflammable. The best containers are aluminium fuel bottles with screw-on lids. Mark fuel bottles clearly (add a warning symbol) to avoid confusing them with water bottles. Use a permanent marker or paste the label on the container and cover with transparent plastic.

WARM FOOD AND DRINK RAISES THE BODY'S CORE TEMPERATURE.

AN ALUMINIUM SET COMPRISING GAS BURNER, STOVE AND FLASK.

Knots

Practise knots and lashings until you can do them easily and swiftly.

Sealing nylon ropes

To prevent the nylon from unravelling, burn the rope ends. These can also be sealed by pressing or rolling the ends with a very hot piece of metal.

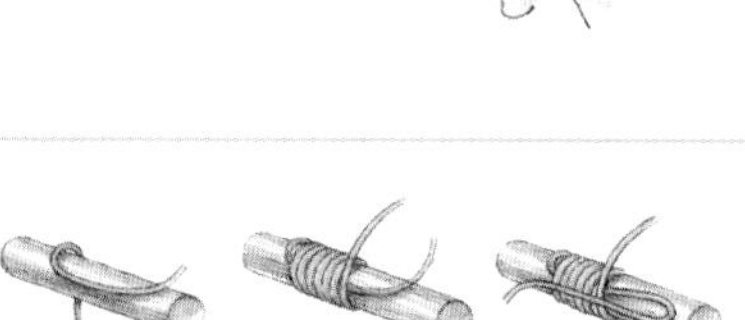

Whipping

This method is used to prevent the ends of multi-core ropes from unravelling, and on sisal and fibre ropes which cannot be heat-fused. Whipping, where thin cord is wrapped around the thicker rope, creates added grip to axe handles, knives or the handles of stretchers which need to be carried over distance.

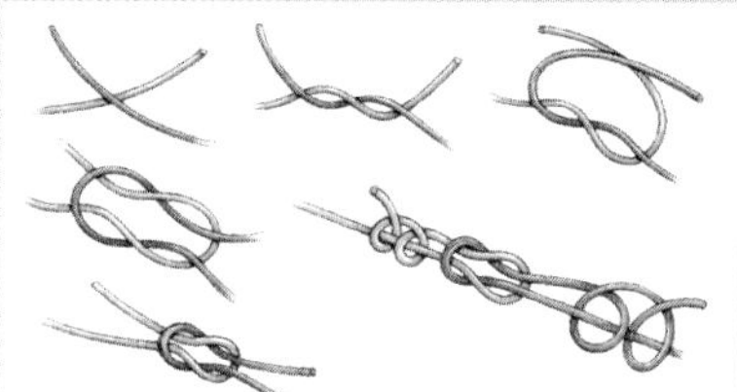

Reef knot with half-hitches

One of the most common knots; it is used to tie ropes of similar thickness together. Ties fairly easily. Safer when finished with a half-hitch or two on either side.

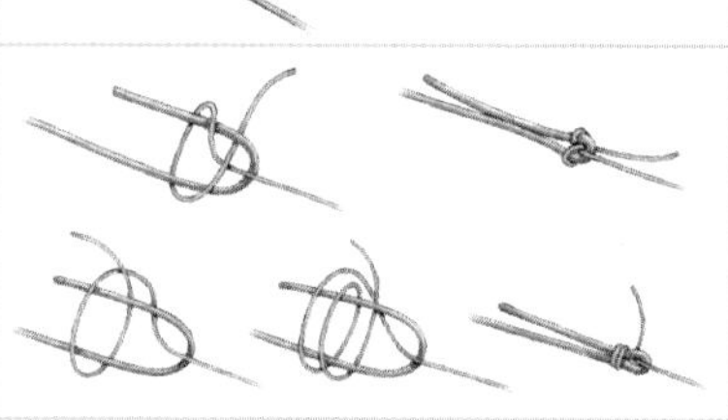

Sheet bend; double sheet bend

Ties two rope ends of differing thicknesses together. Unties easily. Double sheet bend is safer, but more complicated.

Overhand loop

A quick and easy knot that holds well under load, it can be tied at the end or middle of a rope. Useful for making nets or joining rope ends.

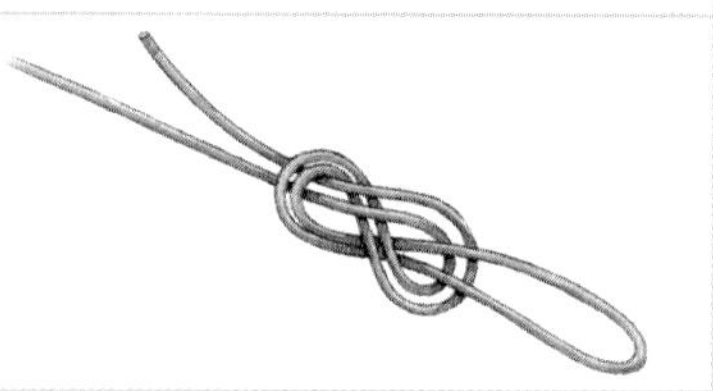

Figure of eight

This knot is tied quickly and is less likely to slip. Climbers use it to attach a rope to a carabiner fixed on a climbing harness. Stronger than overhand loop.

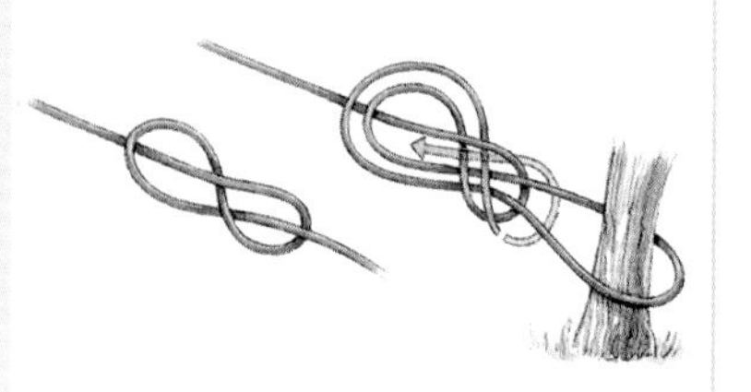

Rewoven figure of eight

Used to make a loop around an anchor point; also, to tie rope directly onto a climber's harness. The two ends lie constantly parallel and emerge at the same end.

Round turn and two half-hitches

Secures a line to a post. Easily tied, even when the rope is under tension.

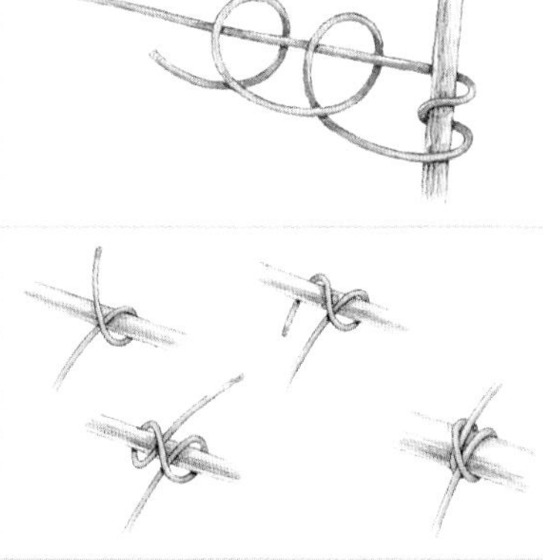

Clove hitch

Good for starting or ending lashings; both ends can take strain. Can be used in the middle of a rope if the rope is slipped around the end of a spar or beam.

Timber hitch

Used in lashings or to attach a beam or log that is being dragged. An extra half-hitch can be added to hold heavy spars.

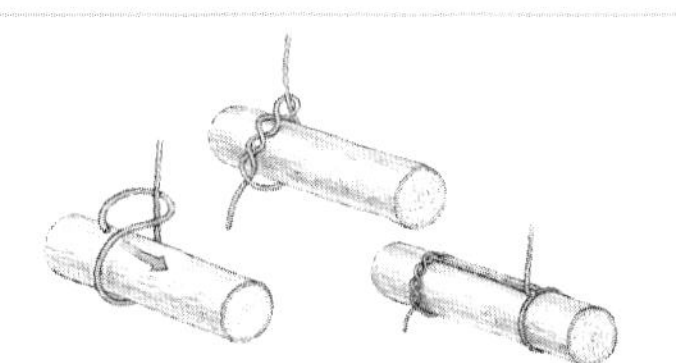

Lashings

Square lashing

Used wherever poles cross at right angles, e.g. rafts and stretchers.

- Start with a clove hitch or timber hitch under the crosspiece.
- Take the rope several times over and under the crosspiece, pulling tightly.
- Make a full turn around the horizontal pole so you can wrap in the opposite direction for four turns. Secure with a few half-hitches or a clove hitch.

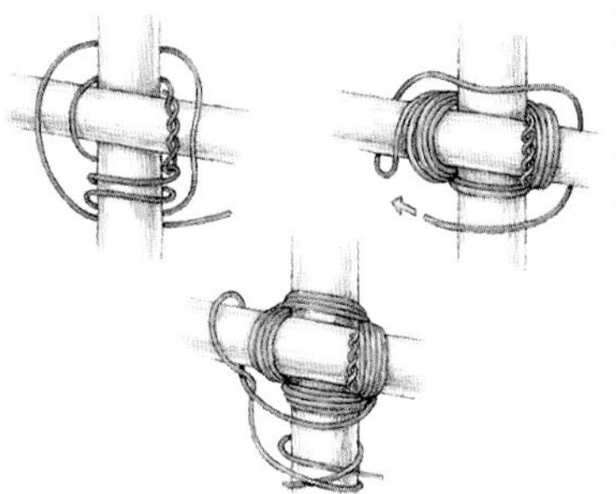

Shear (diagonal) lashing

Joins poles that meet on the diagonal, or any angle other than a right angle. Used in tepees, A-frame bridges or shelters.

- Tie a clove hitch on one spar, then do a few loose horizontal turns around both spars. Loop several vertical frapping (tightening) turns between spars.
- Finish with a clove hitch on the second spar.

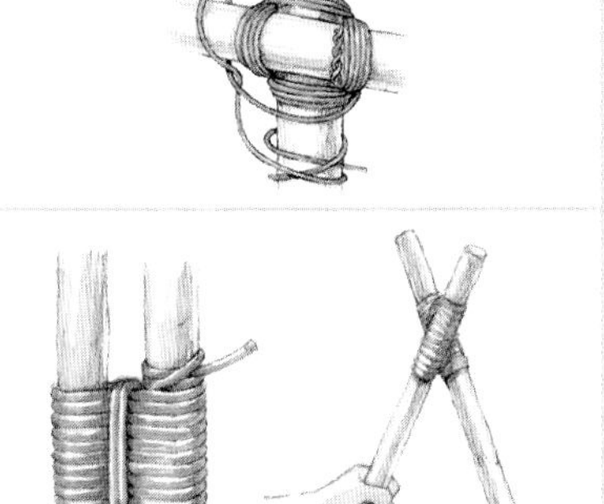

Improvised cord and rope

Vehicles have material in the form of electric wire, cord in carpets, seat covers, or nylon bands in some tyres. You can also use vines, grass, the bark of some trees, leaves and animal hair to make cord of varying strengths. Once you have plaited several lengths (you need to keep the plaits tight when you work with natural fibres), plait three pieces tightly together in order to make a much stronger rope.

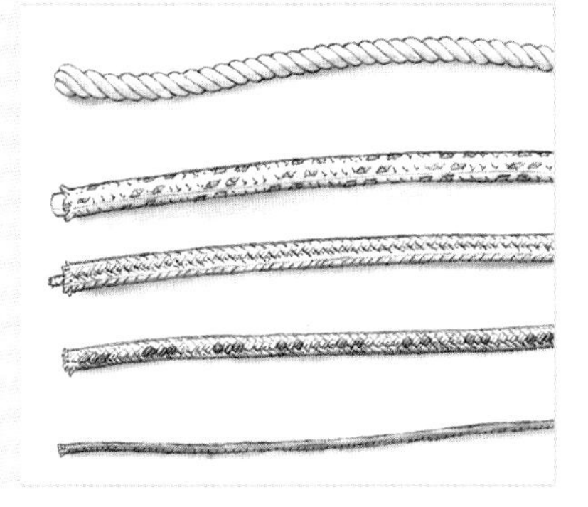

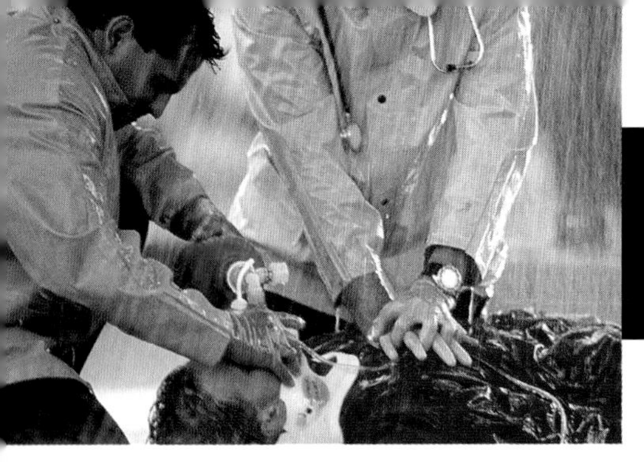

First Aid

Whether you are on a gentle hike in the country or a multi-day mountain trek, if you can no longer access emergency medical services it is up to you to manage any medical situation that may occur. How you respond and what you do can make a major difference to the outcome. Few of us have adequate emergency training. If you are a serious wilderness enthusiast, it is advisable to seek out special wilderness medicine courses for training in emergencies; if you are a casual adventurer, attending a hands-on first-aid or basic life support course is worth it. Keep your qualifications up to date.

Prepare yourself before setting out by considering what the likely hazards are. Does any member of the group have a medical condition that could become a problem? How would you access emergency services – is there cellular phone coverage in the area, and what are the numbers? Establish if any members of the group have medical or first-aid training and consider what sort of medical equipment you need to take along (see first-aid kits pp84–85).

Medical emergencies

These always happen when you least expect it. Suddenly someone falls, a vehicle overturns or a lifeless body is pulled out of the water. The situation feels unreal, you have a paralyzing moment of panic. How you react and the actions you take can make all the difference in a medical emergency.

It helps to simply state what has happened as it will focus the rest of the group. If you are the leader, your primary responsibility is everyone's safety. The leader should preferably not be directly involved with patient care unless he/she is the only medically competent person. The leader needs to keep an overall eye on the situation and delegate tasks. Note that the noisiest patient is not necessarily the worst off!

opposite RESCUE OPERATIONS OFTEN INVOLVE COMPLICATED, TIME-CONSUMING PROCEDURES WHICH DEMAND MUCH ENERGY.

Emergency care protocol

Hazard

In other words, protect yourself, other group members and the patient from further harm.

- If, for example, you are on steep ground, first secure the patient and rescuer with a rope.
- Only in an extremely hazardous situation, such as a fire, should the patient be moved without first carrying out a proper medical assessment.
- If available, don latex gloves before touching body fluids, although the major risks – HIV and Hepatitis B virus – cannot penetrate intact skin.
- Safeguard any sharp objects that are contaminated with blood (glass shards, injection needles).

Hello

In other words, talk to the patient.

- Is the patient conscious? Do you have permission to treat him or her?
- If he/she answers back you know that the brain is functioning and that immediate CPR is unnecessary.
- Immediate verbal reassurance is important.

Help

In other words, access emergency services.

- Call people around you to come and help if they are not yet aware of the emergency.
- Decide whether to call for outside help before starting emergency treatment (in a wilderness area this may not be feasible).

Basic first aid

ABC Principles

Perform a primary survey to check for life-threatening conditions that require immediate attention.

- **Airway:** Ensure the airway is open.
- **Breathing:** If patient is not breathing, give artificial ventilation.
- **Circulation:** Stop any bleeding. If the heart has stopped, perform heart massage.

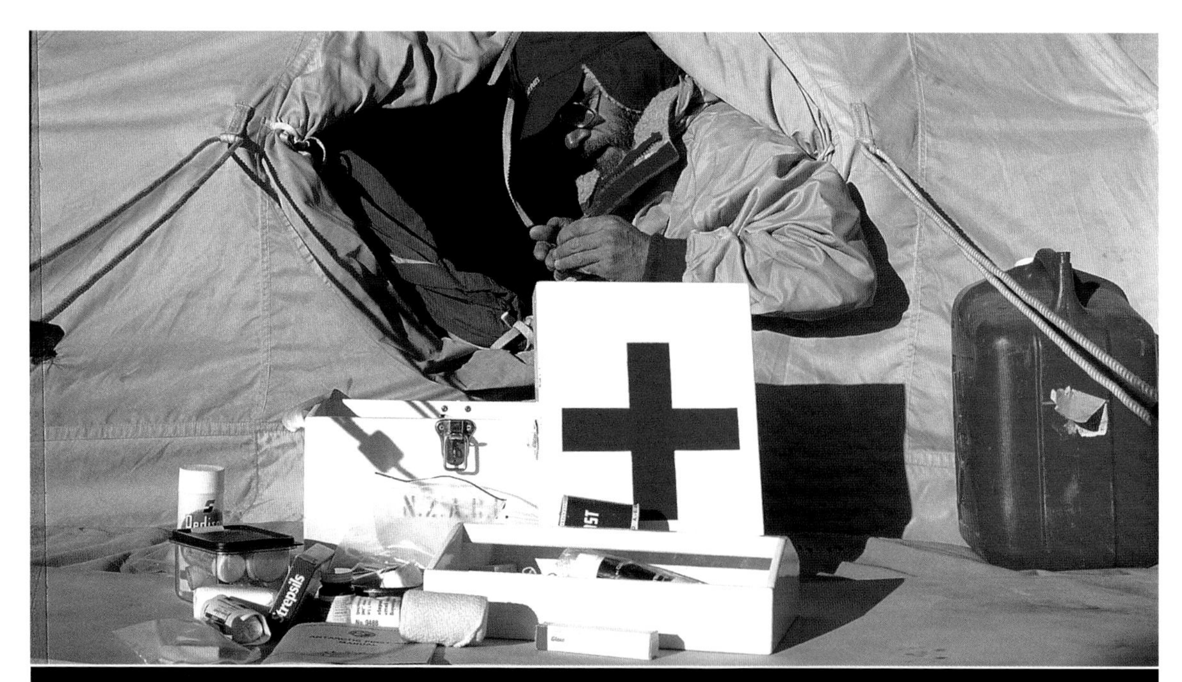

A COMPREHENSIVE FIRST AID KIT IS NECESSARY FOR MAJOR EXPEDITIONS TO WILD AREAS WITH A FAIR-SIZED GROUP.

Airway

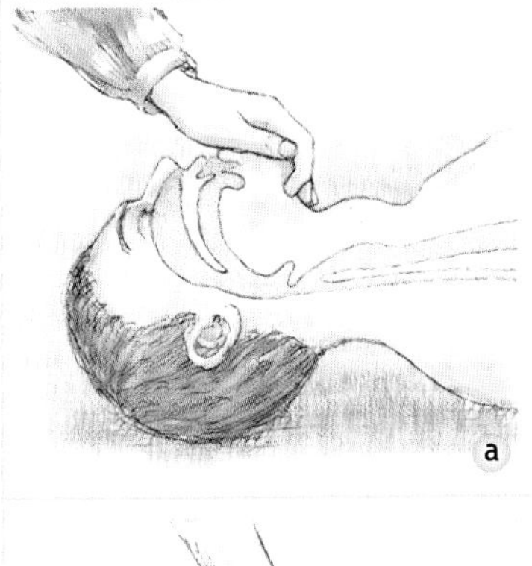

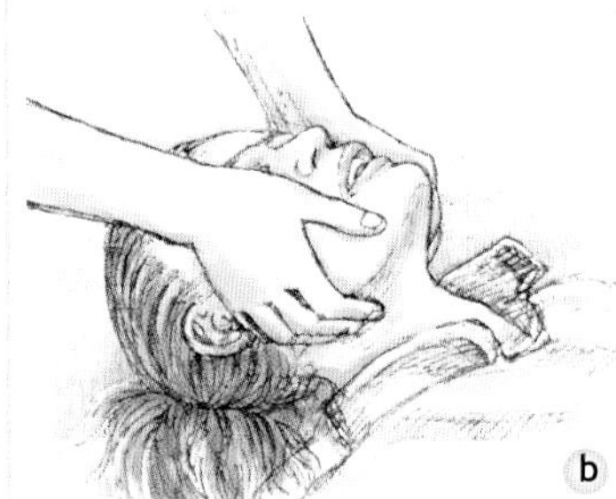

Blockages of the airway

In a conscious patient, airway blockage can result from severe facial injuries, insect bites, swelling from infection, inhalation of hot gases or a solid piece of food that 'goes the wrong way' and then gets stuck in the voice box. The patient will struggle to breathe in, and will snore, gurgle or make a high-pitched sound. There may also be sucking in of the ribcage and throat.

In an unconscious patient, open the airway by pushing the lower jawbone forward (see [a]) or applying pressure behind both angles of the jaw (see [b]). The neck should be held in a neutral position in case of a neck fracture. You may have to put your finger in the mouth to remove any obstruction (e.g. lumps of food or false teeth).

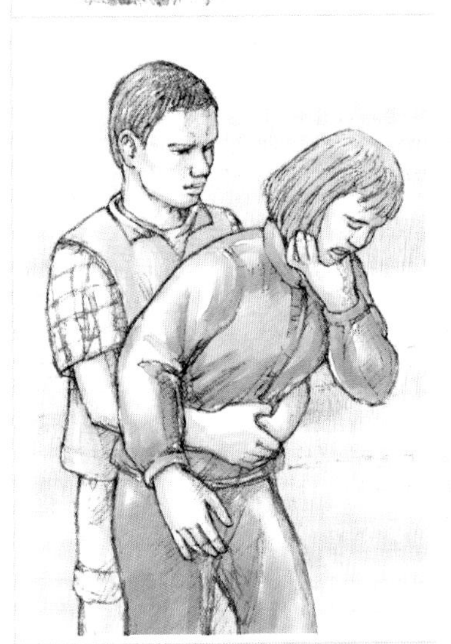

Heimlich manoeuvre

If an adult gets a solid piece of food stuck in the voice box, the Heimlich manoeuvre (abdominal thrust) can be used to dislodge the obstruction.

Stand behind the patient and wrap your arms around his waist. Make a fist with one hand, cover it with the other and give a hard, swift upward thrust just below the ribcage while compressing the ribs with your arms. Repeat till the food has been dislodged.

Small children should first have their mouths checked for removable objects, then held face down and thumped between the shoulder blades to dislodge the obstruction.

Recovery position

Unconscious patients should be turned onto their side in the recovery position (see left). Remember to protect the neck and spine while moving an injured patient into this position. Keep the patient in the recovery position even during transport; it ensures that gravity keeps the airway open and any vomit can run out of the mouth without entering the windpipe.

Breathing

Breathing problems can result from injuries to the chest, accumulation of fluid in the lungs caused by altitude and a near-drowning incident, pneumonia or an acute asthma attack. The normal resting breath rate for an adult is 15 to 25 breaths per minute. More than 30 breaths per minute may be an indication of a breathing problem. Always assume that breathing difficulties are serious and get help promptly. Let the patient sit up if he/she is more comfortable that way. Give oxygen if available.

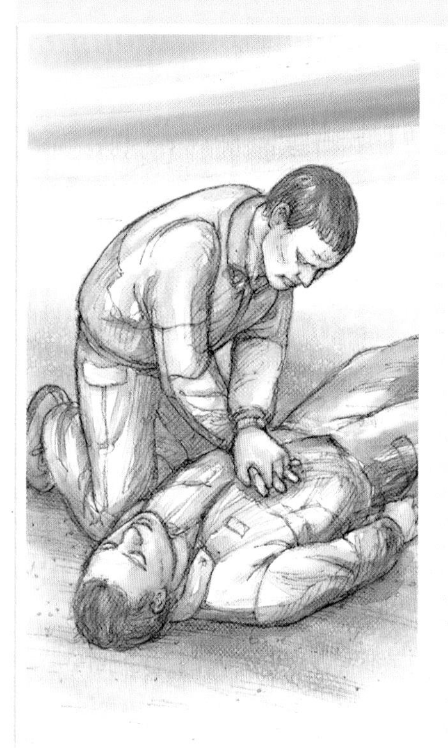

CPR

Cardiopulmonary resuscitation (CPR) is best learnt on a training course but the following is a brief guideline. Perform CPR promptly and proficiently, but you need to recognize that CPR offers only a slight chance of survival to a patient with cardiac arrest. In the wilderness, situations where CPR may save a life are a stopped heart from hypothermia, near-drowning and lightning strike.

The patient should be lying on his back. Kneel at his side and place the heel of your hand in the centre of the lower half of his chest, with your other hand resting on the first (see left). Thrust down with straight arms, depressing the chest by 4–5cm (2in). Pause after 15 compressions, and give two breaths (over two seconds each) of mouth-to-mouth respiration. Do 100 compressions per minute. After four cycles, check for breathing and a pulse.

Survival after more than 15 minutes of CPR is very unlikely. Except in cases of hypothermia, it would be acceptable to stop after 30 minutes of CPR if there has been no response.

Artificial respiration

Remove any obstructions, then lift the patient's chin to open the airway, seal his/her mouth and nose with your mouth (see left) and breathe deeply for two full seconds. Check that the patient's chest rises with your exhalation. Administer 10 to 12 breaths per minute.

Circulation

The next most important priority is to stop any bleeding from an open wound.

- Control by applying firm direct pressure on the area with any reasonably clean piece of cloth, or your hand (or the patient's).
- To stop arterial bleeding, press very hard on the dressing for at least three minutes, then maintain the pressure with a bandage.
- If dressing and bandage become soaked with blood, apply a second dressing over the first.
- Elevate the bleeding zone; check there are no constrictions above it (e.g. a rolled-up trouser leg) which will encourage venous bleeding.
- Only use tourniquets as a last resort to stop bleeding after a traumatic amputation, e.g. loss of a limb from a shark attack.

In advanced blood-loss shock, the patient becomes confused and loses consciousness. Keep the patient still while he or she is lying flat and elevate the legs; this returns some blood to the central circulation. Rough movement or transportation of a shocked patient can aggravate the bleeding and cause rapid deterioration in the patient's condition.

Internal bleeding can occur from blunt or penetrating injuries to the chest and abdomen. There is no way of controlling this type of bleeding except with surgery. You can only treat for shock.

Below In a bleeding patient it is important to conserve blood; elevating the area will encourage the blood to return to the body's circulatory system.

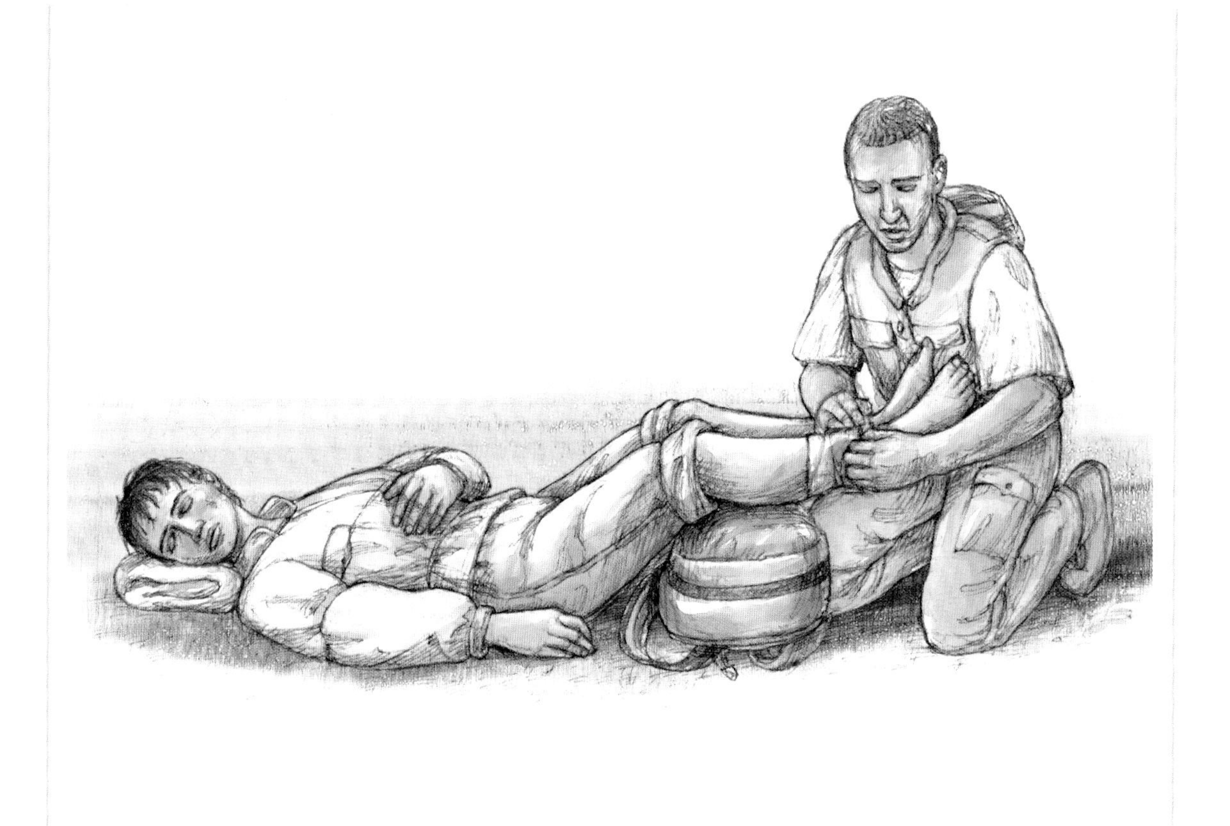

Spinal stabilization

If the patient has been injured in a fall or vehicle accident, automatically assume a spinal (neck or back) injury and make sure that neck and spine are stabilized. The patient should lie still with the neck in a neutral position – not twisted, bent forward, backward or sideways. Unless it is essential to move the patient to avoid immediate, life-threatening danger, only move the patient after you have ruled out a neck or spinal injury; or after you have immobilized the neck with a rigid collar and the spine on a stiff board or stretcher.

Proceed as if all unconscious, injured patients have a spinal injury. Conscious patients who have no pain in the neck or spine and no numbness or paralysis of the limbs probably do not have a spinal injury. Any period of unconsciousness should be regarded as very serious as this could indicate brain injury.

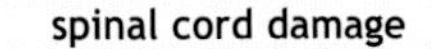

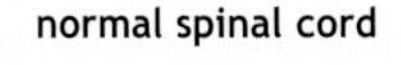

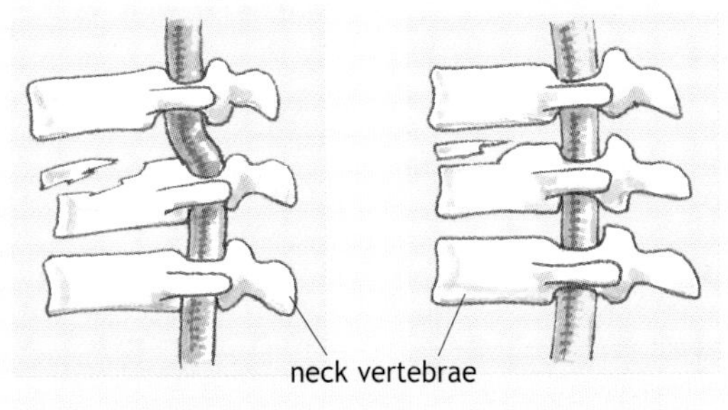

Fractures

The symptoms of a fracture, namely intense pain, inability to move the limb and deformity, are usually obvious. Broken bones are very painful but seldom life-threatening; however, multiple fractures or those involving the femur (thigh bone) can cause enough blood loss to result in shock and death. Open fractures (when a wound overlies the fracture) and those with damaged nerves and arteries are more serious.

Fractures must be splinted to reduce pain and prevent further tissue injury and blood loss from movement. You can splint a leg to the opposite leg and an arm to the chest. Any suitable straight, rigid object such as an ice axe or branch can be adapted as a splint. Rolled up thin (dense) foam mattresses work well. Make sure the splint is well padded and always check the circulation after applying the splint.

In the case of a rib fracture to the chest, treat by giving pain-relieving medication. The patient should sit up to help the breathing process.

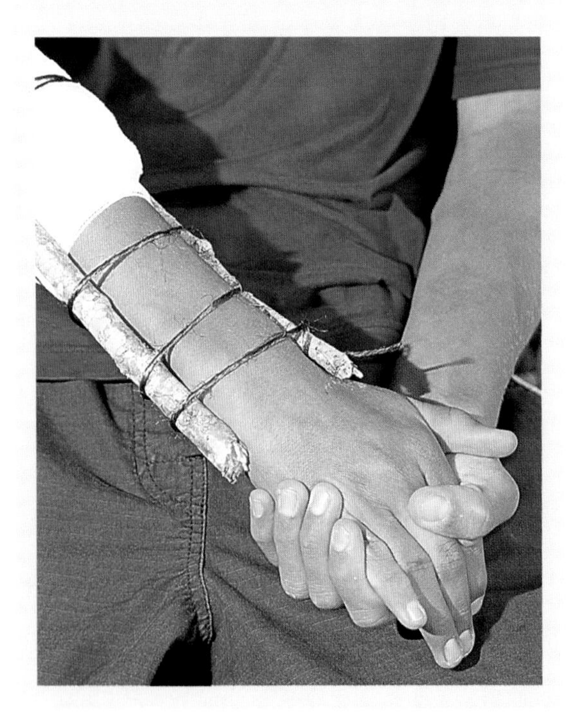

Burns

Cool the burned area by immersing it in clean water or pouring water on it. The immersion can be maintained as long as is practical, as it helps pain and limits tissue destruction.

Cover the burn with a sterile dressing or a clean cloth. Do not try to clean the burn and do not apply ointments. Burns that have penetrated the full thickness of the skin, resulting in destruction of the tissue, will need skin grafting later.

Large burns (involving more than 20 per cent of the body surface area) can cause enough plasma loss to result in shock to the body. Urgent hospitalization is required for these and other burns on the face and hands.

Sprains

These are injuries to the soft tissues surrounding joints where the muscles, ligaments and tendons are stretched or torn. Sprains will normally recover given time. The most common sprain is an inward twisting of the ankle. Use the RICE principle: Rest, Ice, Compression and Elevation. Stop walking, rest and apply ice, snow, or cold water if this is available.

Strap a sprained ankle with broad, non-stretchable tape to hold the ankle in the opposite way to the direction of injury. This should be followed by a firm, elastic bandage applied over the tape to further support the sprained limb. Elevate the limb to help reduce swelling.

If it is essential to continue walking, it may be better not to remove the boot after an ankle sprain, as swelling may prevent you getting the boot back on.

Dislocations

Dislocations are serious joint injuries that cause bone displacement. They are very painful and the affected limb cannot be used. In hip and shoulder dislocations, the long bone literally comes out of the joint socket. The limb's nerve and blood supply may be affected and will require urgent medical attention.

Dislocations can often be corrected by firmly pulling the limb while pushing the bone back into the joint, immediately after an injury if possible. This also helps to ease the pain. This is not an easy process, and fractures can occur or worsen if done incorrectly, but in a survival situation you may need to attempt it. Otherwise, support the dislocated limb in a splint.

Below Here, a dislocated shoulder is treated by applying steady traction on the arm using a helmet weighted with a rock.

Heat-related conditions

Minor heat-related conditions include heat rash, heat edema (feet swelling), heat cramps and heat syncope (a faint that occurs after stopping vigorous exercise in a hot environment).

Heat exhaustion – This is the most common of the more serious heat-related conditions. Symptoms include exhaustion, dizziness, mild mental changes, nausea and headache. The patient may pant and sweat profusely. Treat all heat conditions by resting in the shade, rehydrating and removing excess clothing.

Heat stroke – This is a rare but serious condition that can be fatal. The body temperature (normal 37°C, or 98.6°F) can rise as high as 46°C (115°F). The main symptoms of heat stroke are confusion, seizures and coma. Some patients may not sweat despite the heat, although others sweat profusely.

Treat by aggressive cooling, fanning and using all available water. Keep the patient in the recovery position, watch the airway and arrange for urgent evacuation to a hospital if possible.

Cold-related conditions

Hypothermia

Hypothermia (low body temperature) occurs when the body fails to conserve heat, causing uncontrollable shivering and mental confusion. Unless heat loss is prevented, the patient's condition will deteriorate, resulting in coma, cardiac arrest and death. It occurs when the core (heart, lungs and brain) temperature falls below 35°C (95°F). In needing to conserve heat, the body reduces the blood supply to the skin and limbs, resulting in blue fingers and toes.

Hypothermia is aggravated by hunger and fatigue, illness and high altitude. Highly at risk are young children, adolescents, thin persons and elderly group members. Injured and immobilized patients can develop hypothermia under surprisingly mild conditions. Look out for signs of altered mental function (stumbling, staggering, confusion, mumbling or uncharacteristic irritability), take action. Stop, seek shelter, rest and rewarm the victim(s).

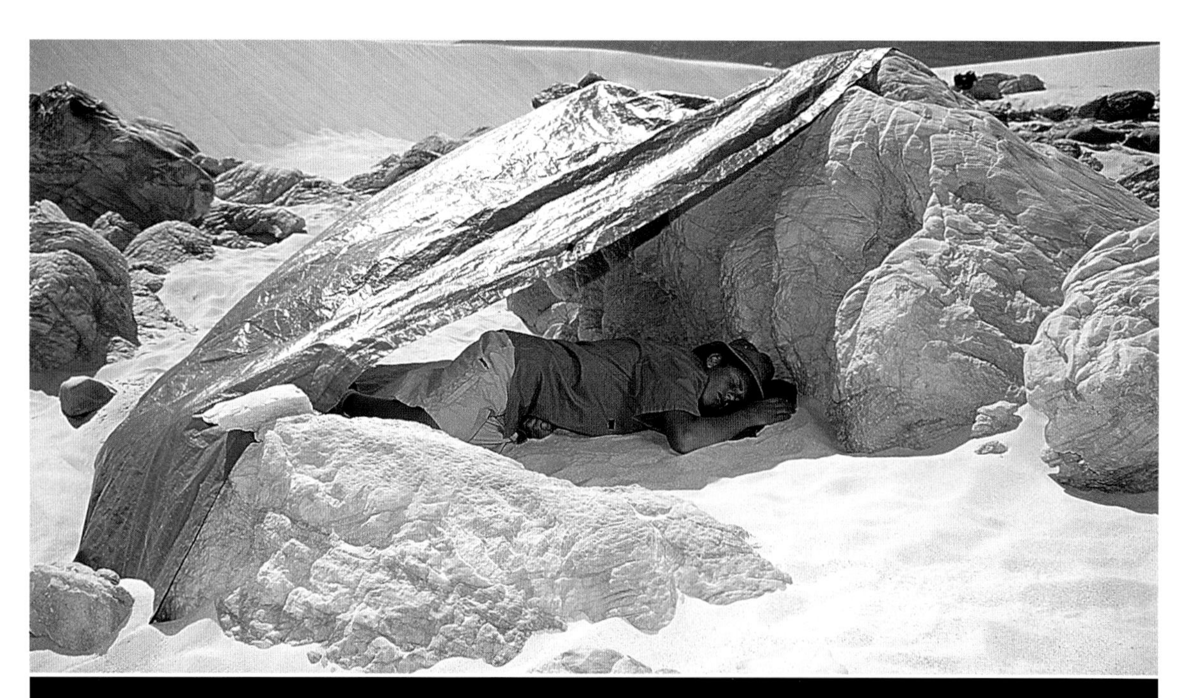

IN HOT DESERT TERRAIN IT IS IMPORTANT TO CONSERVE BODY MOISTURE. SHELTER FROM THE BURNING SUN UNDER A MAKESHIFT TENT WEIGHTED WITH STONES TO PREVENT SERIOUS CONDITIONS SUCH AS HEAT EXHAUSTION AND HEAT STROKE.

Taking action

- Get the victim into dry clothes, then place in a sleeping bag inside a tent, insulated from the ground.
- If dry clothes are not available, use a plastic covering to prevent heat loss by evaporation. Cover the head as it loses heat very rapidly.
- The body heat of companions can warm a patient.
- Avoid hot objects applied to the skin as they can easily cause burning because of the reduced blood supply to the skin.
- Conscious patients should be given plenty of warm liquids to replace urine loss.
- If the patient is unconscious place him/her in the recovery position and monitor the airway.
- Give mouth-to-mouth respiration if breathing ceases; administer CPR if the heart stops. Severe hypothermia mimics brain death; when in doubt, continue with CPR until the patient has been warmed up.

DARK NEGATIVE FILM PROVIDES IMPROVISED EYE PROTECTION AGAINST GLARE FROM DESERT SAND OR SNOWY PEAKS.

Stages of hypothermia

Mild

Core temperature 35°C–32°C (95°F–92°F)

- Complains of severe cold
- Poor judgement, confusion, irritability
- Slurred speech, stumbling
- Uncontrollable shivering
- Cold, blue hands and feet
- Stiff muscles
- High urine production leading to dehydration

Moderate

Core temperature 32°C–28°C (90°F–82.4°F)

- Decreased level of consciousness
- Shivering may stop
- Muscles are stiff and rigid
- Irregular heartbeat

Severe

Core temperature below 28°C (82.2°F)

- Deeply unconscious
- Slow breathing
- Slow, irregular heartbeat
- Heart may stop

Snowbound conditions

Frostbite – occurs when body tissue freezes; it most commonly affects the fingers, nose, ears and toes. Frostbitten areas turn from an initial white via a blistered stage to black and may lead to permanent tissue damage if affected areas are not warmed up.

Snow blindness – this is caused when too much ultraviolet light enters the eye, damaging the cornea. The resulting inflammation and pain often does not manifest itself for 10–12 hours after exposure. To avoid blindness in snow conditions, wear sunglasses with side shields or wraparound lenses. Emergency glasses can be made by cutting slits in cardboard or by using dark negative film.

Near-drowning

Drowning is often preventable. Insist that boaters wear life jackets, that children near water are always supervised, and do not combine alcohol intake with swimming. Submersion victims lose consciousness from lack of oxygen to the brain. Most struggle and inhale water. If the patient has stopped breathing, start the ABC steps outlined on p74 immediately.

Beware of a possible neck injury if the patient has just dived into the water. If the patient is unconscious but breathing, place him/her in the recovery position (he/she is likely to vomit swallowed water) and keep the airway open.

It is worth doing CPR even after long submersion as cold prolongs the time the brain can function without oxygen. Submersion victims are at risk of developing 'secondary drowning' – breathing problems that develop later after inhaling water. Give oxygen and take to hospital as soon as possible.

WHEN RESCUING A MAN OVERBOARD, BE AWARE THAT GULPING DOWN MOUTHFULS OF WATER CAN LEAD TO SECONDARY DROWNING.

Bites and stings

Bites from snakes and insects can be prevented by wearing suitable footwear and long pants, being very cautious where you put your hands and feet, and inspecting your boots before putting them on.

Snakebite

Note that, generally, if you don't bother snakes, they won't bite you. The majority of snakebites occur while attempting to capture and play with snakes – and they are seldom fatal.

The most dangerous are snakes that produce neurotoxins, which cause muscle paralysis. They are mainly Cobras of Africa and Asia, Mambas of Africa and Coral snakes of North America. Adders, such as Africa's Puff Adder, produce digesting toxins that cause severe tissue damage. The venomous Rattlesnakes of North America cause both paralysis and tissue damage with their toxin.

In the case of a snakebite, calm the patient, who should avoid any movement which will spread the toxin. Apply a firm crepe dressing to the affected area and evacuate the patient by helicopter if possible. If there are any signs of envenomation, snakebite serum can be effective. Suctioning the wound with a suction pump may be effective if immediate. Applying a tourniquet will probably do more harm than good. Watch for signs of respiratory depression and apply mouth-to-mouth respiration if necessary.

Spider and scorpion bites

Few spiderbites are more than a nuisance, bar the Black Widow and Brown Widow of Europe, North America and Africa, and the Funnel Web Spider of Australia. Africa's Violin Spider can cause severe tissue damage – similar to a Puff Adder – but is seldom (if ever) fatal. Scorpion stings can be excrutiatingly painful although seldom fatal – ice applied can help. Multiple stings can produce toxins that affect the nervous system, causing writhing and jittery movements. Treatment is as for snakebite. Children with a smaller body mass are more at risk.

Bee, wasp and hornet stings

Wasps, hornets and particularly bees are responsible for more deaths than snakes, spiders and scorpions combined (especially the African Honey Bee). A single bee sting is only a problem when a patient is hypersensitized because of previous stings or a natural allergy. Antihistamine medication (injected or in tablet form) will help. Multiple bee stings can cause shock, respiratory depression and airway obstruction from swelling. After a severe attack, lie the victim down, remove the stings and apply CPR if needed. Fast removal is vital – bee stings continue to pump poison into the body for up to 20 minutes after an attack. Remove stings by scraping them off with a sharp blade or a needle. Never squeeze or handle the sting; this causes more poison to enter the system.

Marine creatures

These can be divided into those that sting, such as jellyfish, bluebottles and fire coral, and those that 'stick' such as Stone Fish, Barbel and Scorpion Fish (see above). The pain from 'stingers' can usually be relieved by applying vinegar or alcohol to the sting. Pain from the 'stickers' may be intense and is relieved by immediately submerging the affected part (usually the foot) in water as hot as you can tolerate.

Travel and tropical infections

Travellers to malaria zones should get expert advice and take the requisite prophylactics. Mosquitoes also transmit dengue (viral infection) and yellow fever. Make sure you have the necessary inoculations or vaccines required by various countries.

Hepatitis B and HIV are transmitted by using unsterilized needles and syringes, through blood transfusion and sexual intercourse. An effective hepatitis-B vaccine is available and is recommended.

First aid kit

First aid or medical training is important; it is no good taking along equipment and drugs that no one in the group is qualified or experienced to use. Select carefully, always considering weight and space when packing a first aid kit. Choose items that are multifunctional and keep them all together in a waterproof container. It is important to bear in mind that the first aid kit is for emergencies only.

Medical kit for a small group

(small hiking group travelling for about a week to a remote location)

- 2 pairs disposable latex gloves (protection from blood)
- Pocket mask (to provide protection during mouth-to-mouth respiration)
- 1 large (300 x 300mm/12 x 12in) sterile wound dressing (cover wounds; stop bleeding)
- 2 small (100 x 100mm/4 x 4in) sterile dressing (bleeding & wounds)
- 2 burn dressings, e.g. hydrogel (200 x 200mm/8 x 8in)
- 1 crepe bandage (±100mm/4in wide) (joint sprains; hold dressings in place)
- 1 small roll adhesive tape
- 1 SAM® splint (fractures)
- Space blanket
- 2 sachets povidine-iodine disinfectant (10ml/⅓oz)
- Low-reading clinical thermometer (diagnosing fever and hypothermia)
- Scissors and tweezers
- Needle (remove splinters)

Medication (small quantities):

- Paracetamol/codeine tablets for pain (Acetiminophen in the USA)
- Anti-inflammatory drugs, e.g. Ibuprofen or aspirin (sprains)
- Antihistamine tablets, e.g. Promethazine (allergy; nausea)
- Oil of cloves (toothache)
- Loperamide (diarrhoea)
- Antibiotic, e.g. Ciprofloxacin (infection)
- Antifungal ointment
- Saline eye wash (.9 per cent) — 10ml sachet

Mini medical kit

This personal mini survival kit also includes, besides the basic essentials such as compass, flint and waterproof matches, a mini medical kit. It is devised in such a way that it will fit into a small tin, yet contains a surgical blade, safety pins, adhesive plasters, butterfly sutures, and needle and thread.

Essential information for the rescue team

- Nature and causes of incident (e.g. injury, illness)
- Exact location
  - GPS position, map coordinates
  - distance and direction from identifiable feature
  - description of landform/special features, e.g. river bend
- Number, names and ages of patients
- Medical condition of each patient
  - vital signs
  - level of consciousness (AVPU)

A : Alert

V : responds to Verbal stimuli (i.e. talking)

P : responds only to Painful stimuli (e.g. pin prick)

U : Unresponsive

  - medical expertise in group
  - injuries
  - treatment applied
- Local weather conditions
- Local access difficulties (e.g. on a cliff face)
- Site of helicopter landing zone
- Number of uninjured members in the group
- Group situation
  - shelter
  - food
  - medical supplies
- Signalling methods

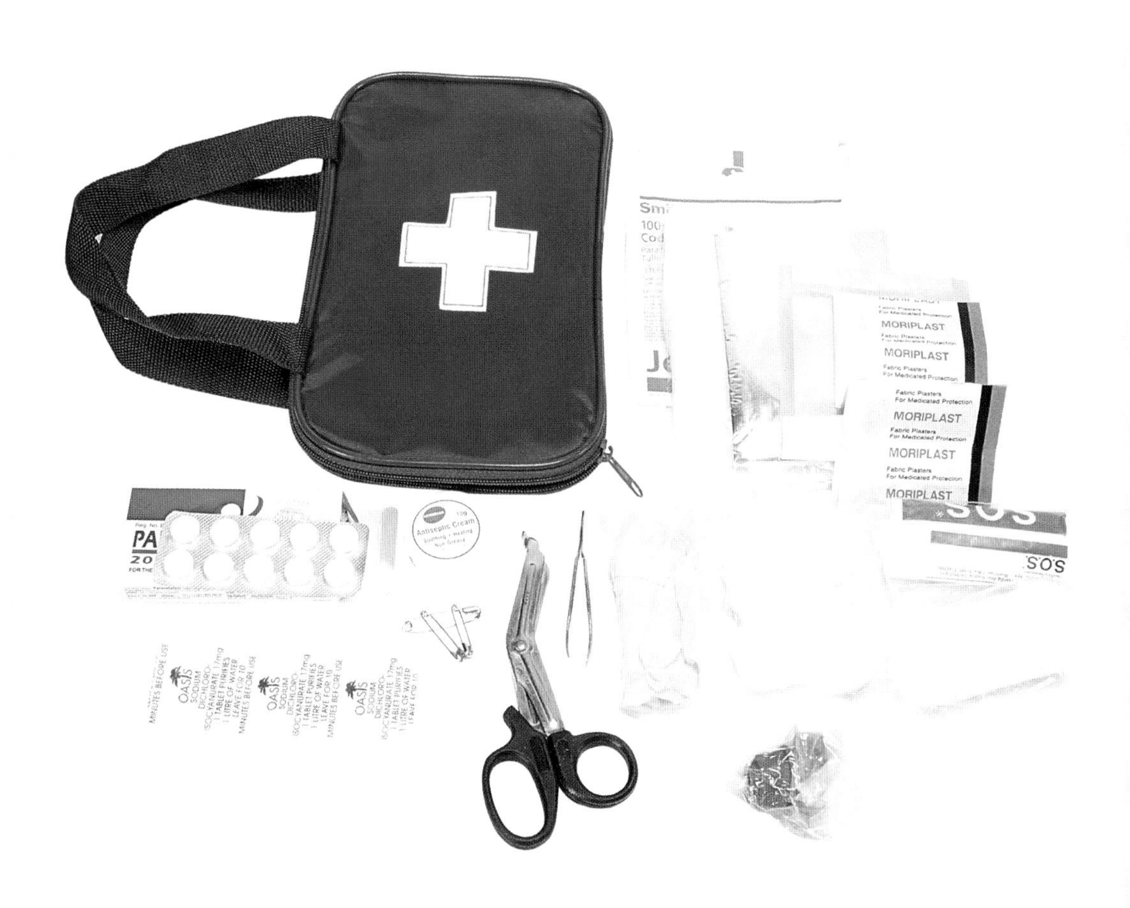

Survival Disasters

Any form of trip passing through or over wilderness could turn into a disaster or survival situation as a result of a shipwreck, the crash of an airplane or other vehicle, or a natural disaster. The ability to calmly assess, evaluate, prioritize and plan amid the chaos and confusion remains a vital component in a survival situation.

Action checklist

Injuries Are you personally injured? Can your injuries be addressed later?

Immediate dangers Do unstable wreckage, volatile fuel, noxious gases, smoke, dropaways or deep chasms pose a danger to anyone? Are people aware of this, and able to help themselves? Will moving them put you and others at risk? How far away is a 'safe' distance?

Treatment priorities Should the seriously injured be left until later for the sake of those who are still fit and have only minor injuries?

Special needs Are there any babies, children, elderly or disabled individuals who need assistance? Are you able to help? Are there others who can assist you?

Resources Check around for useful items – food, water, clothing, shelter, radios, batteries, lamps, fire extinguishers, rafts, or floatable debris.

Distress signals Are you able to send a distress signal immediately (i.e. radio, flares, transponders on boats)?

Staying together People often wander off mindlessly; ask stragglers to perform simple but manageable tasks (e.g. checking that everyone's life jacket is fastened, counting children in the group).

opposite THIS YACHT DUMPED UNCEREMONIOUSLY INLAND IS TESTIMONY TO THE DESTRUCTION THAT CAN BE CAUSED BY THE POWERFUL WINDS OF A HURRICANE (SEE P93).

Safeguard resources and supplies Commandeer food, drink, matches, lighters, lamps and communication devices on behalf of the entire group.

Rationing Adopt a 'worst case' scenario attitude and start rationing food and water immediately.

Shipwrecks or sinking craft

Small craft Every person should have a life jacket or some good flotation aid. Do not overload a boat. When in the craft, keep people positioned as low as possible to aid stability. If a distress signal has been fired, let down a sea anchor to reduce drift of the life boat.

Larger boats Beforehand, take note of all exits belowdeck and the position of lifeboats. In an emergency try to grab and put on as much clothing as possible to provide insulation in colder waters. A life jacket is a priority; also grab anything that could aid flotation in the water – empty plastic bottles or tins, expanded polystyrene foam, inner tubes or wood.

Get as close to the water as possible before jumping, especially if your life jacket is already inflated. Make sure you swim away from a vessel that is likely to sink to avoid the undertow.

Adrift with a sports craft

Divers, boardsailors, sea canoeists, body-boarders or surfers should stay with their craft, even if it is partially wrecked or overturned. It is better to stay on the craft. Link your hands underneath the board, and do not detach your safety leash.

Sea canoeists and boardsailors should carry a waterproof whistle as well as a flare pen or strobe light fastened to the craft with strong tape or cord.

Visibility to searchers is very important. If you hear or see a search vessel, try to raise your hand high above the water and wave a colourful item (e.g. socks, underwear or swimming costume).

Adrift without a craft

If you fall off a small craft or get carried out to sea while swimming, save energy and lie in the water. In smooth water, you can float on your back. If the water is rough, then try 'drownproofing': lie on your stomach with arms extended and your face in the water. Raise your head every now and again to breathe. Alternate by treading water, then lying relaxed in the water. If in a group, huddle together to preserve body warmth.

Makeshift flotation devices can be created by removing long pants or a shirt, and securely tying the ends of the legs or sleeves in a knot. Wave the garment over your head to fill it with air, then gather the bottom and tie it off with a belt, shoelace, or hold the end beneath the water to create air sacs. Place the air-filled garment around your neck.

River rescue

- Apply the golden rule of rescue: Reach, Throw, Row, Go (Rethrog).
- First try to rescue the person with a paddle or stick, next try hurling a throwline or rope. The person in the water should lie on his or her back, grab the line with both hands and bend the head forward slightly to aid breathing. Don't fasten a throwline to a belayer on the banks or a rescuer in the water; the strong current may drag them under.
- If this is still unsuccessful, attempt to row out towards the swimmer.
- The last option is to swim, but only if you are a competent swimmer, have rope assistance and conditions are safe for swimming.
- Do not attempt a river rescue if you are not skilled; rather call the trip leader or river guide.

Wild rivers & whitewater

Rivers once considered the domain of elite experts are now being run by large numbers of amateur paddlers, with an increase in river incidents. Don't be fooled by a seemingly placid river and don't run rivers that appear to be beyond your capabilities.

Adopt the following defensive swimming techniques down a fast-flowing river.

- If you are holding onto your boat, get onto the upstream side to avoid being trapped between it and rocks or obstacles.
- If the craft appears to constitute a hazard, let it go.
- Float on your back with the current, legs in front, and use your arms to push away obstacles.
- If in very large waves, try to swim aggressively away from them to avoid being tumbled. Swim across the flow of the current.
- If you find yourself heading for a strainer (i.e. a partially submerged tree or log), swap position from feet-first to head-first. This will help you to swim or clamber over the obstacle and avoid being trapped by the onrushing water.

right MANY PEOPLE ENJOY WHITEWATER RAFTING – BUT WILD WATER SHOULD NEVER BE UNDERESTIMATED.

Mountain terrain

Often wilderness and outdoor adventure experiences involve mountainous terrain, and many small and large aeroplane crashes occur in mountains. High winds, cold, avalanches and steep mountain slopes all constitute hazards, particularly if members are stressed and exhausted. Only tackle these conditions if you have no other choice. It may well be more risky to climb than to wait for rescue, although this can be a difficult decision.

Tips for taking action

- If you are lost, do not retrace your steps unless you know for certain where you came from.
- Seek shelter as soon as possible but try to stay close to your pathway. If you shelter in a cave or overhang, leave a clear, visible trail for searchers.
- Create markers (arrows out of stones or sticks bent to point in your direction of travel).
- If it is snowing, build high cairns of rocks or stick tripods and attach pieces of cloth, plastic or similar.
- Wear clothing that will protect against hypothermia.
- If you decide to keep moving, progressing to higher ground can help to orientate you through the location of natural features. Alternatively, move downhill and find a stream you can follow.

Lost in Caves

As soon as you know you have lost your way in a cave, you should:

- Preserve any light you have; turn off all light sources.
- Sit facing the direction in which you were travelling.
- Evaluate your situation; establish who knows where you are and when they are likely to start searching. Should you move, or stay put?
- List your resources in terms of food, light, clothes and water.
- Let group members hold hands, touch one another or sit in contact to ensure that they do not split up, and to give security and preserve body warmth.
- If all lights fail, let your eyes adjust fully. This could take 30 minutes or more. When your eyes have adjusted, you can use the dim light of a cellphone or watch to look around.
- If your supply of matches is limited, try lighting strips of cloth wound around a stick, pen or torch.
- A candle rubbed vigorously onto cloth makes it burn more readily. (You could also try lipstick, lip salve, butter or cheese.)
- Crawling is often the safest option. Use socks or wrap some cloth around hands and knees to protect them.
- As you move along, scratch your initials, time and direction of travel into the ground, or build a pile of rocks to assist rescuers – or help you in retracing your steps.
- In very narrow passages, position the largest person in the middle of the group. If he or she gets stuck, team members will be able to assist.
- The end (or lead) person should always have one hand in contact with the cave wall to prevent the group from doubling back on their tracks.

left IF YOU ARE LOST IN UNFAMILIAR MOUNTAIN TERRAIN, WALK CAREFULLY TO AVOID FALLING INTO CREVASSES OR OVER A PRECIPICE.

Environmental hazards

AVALANCHE VICTIMS TRAPPED UNDER 10M (30FT) OF SNOW HAVE SUCCEEDED IN DIGGING THEMSELVES OUT.

Survival techniques

- If caught in an avalanche, try to move to the surface or the side by making vigorous swimming movements with your hands and feet.
- As you feel the avalanche starting to slow down, pummel the snow around you with your hands, legs and arms to create an air space. Lie still; if you can see light, you might be near the surface.
- Listen carefully for sounds of rescue and shout *only* when rescuers are very close.
- If no rescue arrives within a few minutes, then dig upward; allowing saliva to dribble out of your mouth will indicate up and down orientation.
- If you see someone caught in an avalanche, keep watching them as long as possible. When the avalanche stops, move to the point where they were last seen and search rapidly downhill.
- If you have long thin sticks, use them to prod in the snow for the victim(s).
- This initial fast search is vital. 50 per cent of avalanche victims who survive free themselves, 40 per cent are found in hasty searches by their group members, while only 10 per cent are found in later organized searches.

Avoiding an avalanche

Avalanches occur most often after large snowfalls and on convex slopes of between 30–45 degrees. The potential of an avalanche usually increases when snow that has bonded with underlying layers is loosened by light rain or a rapid rise in temperature. Gullies and mountain valleys can be avalanche runnels (pathways); you need to be especially careful when crossing or using them.

Signs to look for are traces of debris below such areas to establish whether they are prone to avalanches. Cracks, large snowballs running down slopes, or noises when crossing an ice-covered slope are common danger signs of an avalanche-prone area. If you must, climb on or below such hazardous slopes in the very early morning or late at night when the snow has frozen. If you have no option but to cross a dangerous-looking slope, send members across singly. You should first loosen your backpack waist belts and straps as a pack or rucksack can get hooked and may trap you under the snow.

As a precaution for areas under heavy snow, include in your kit a Rescue Transceiver (a compact avalanche radio beacon worn on your person which transmits continuous radio signals to assist rescuers in searching and locating lost people or buried avalanche victims). Always ensure that you switch this on to 'transmit' mode when in an avalanche area.

Forest fires

If a camp fire has got out of control or a fire started from sunlight burning on reflective material in very dry vegetation, try to put it out using clothing or a towel. If the fire is already well developed, your most important action is to get out of its direct path and any potential smoke and gas inhalation. Fires generally move faster uphill than downhill. Take note of the prevailing wind. Try to head for a natural fire break such as a large clearing, rocky outcrop or a river.

If water is not present and you cannot avoid the fire, an option is to climb into a deep gully and cover yourself with earth.

You may need to break through the fire. Cover your mouth and face with a cloth (wet it first, if possible). Choose a spot with a thin stand of vegetation and no holes or rocks to trip you, then run fast through the flames. If your clothes or hair catch alight, drop and roll in sand where you can, or use clothing to smother the flames.

If you are trapped in a vehicle that is surrounded by fire, the best course of action may be to remain in it; check that the windows are firmly closed as fire needs oxygen to keep burning. There certainly is a risk that the petrol tank may catch fire and explode, but this seldom occurs in reality.

A BUSHFIRE CAN RAPIDLY CROSS AN IMPRESSIVE DISTANCE; TAKE NOTE OF ITS DIRECTION AND GET OUT OF ITS PATH FAST.

LIGHTNING IS THE RESULT OF STATIC ELECTRICITY CAUSED BY THE INTENSE AGITATION OF WATER DROPLETS WHEN RISING WARM AIR COMES INTO CONTACT WITH COLD AIR; MILLIONS OF VOLTS FLASH BETWEEN WATER PARTICLES DURING THE BOLT'S PATHWAY TOWARDS EARTH.

Lightning

Mountaineers and high-mountain hikers are more likely to encounter this hazard, as lightning usually strikes higher-lying rocky areas. A lightning storm can be detected by a tingling sensation in the skin; the feeling that your hair is standing on end; and the buzzing of metal equipment. The best shelter is in a dry, deep cave, away from any walls and not directly under the cave's overhang. Sit in a hunched-up position, with your feet off the ground and on dry insulating objects such as backpacks or a coiled rope.

Avoid rock fissures and chimneys, especially when wet, as they provide a pathway for lightning discharges. Also stay away from ridges and summits. If you are caught on open ground during a lightning storm, lie flat on the ground with your arms spread out. In this way you will not become the point of discharge if lightning should strike near you.

If someone is struck by lightning, check vital signs and administer CPR immediately. A patient with burn injuries should only be treated later; the priority is to get breathing going and blood circulation to resume.

Hurricanes and tornadoes

A hurricane is a tropical cyclone with wind speeds in excess of 115kph (72mph), measuring force 12 on the Beaufort scale. It brings with it lashing rain, resulting in tidal waves at the coast, is immensely destructive and can measure up to 500km (300 miles) in diameter. Warning signs are alarming variations in atmospheric pressure, sudden enormous swells, banks of cirrus cloud, and unnaturally bright skies at dawn and dusk. After a stormy spell, the eye of the storm passes in a period of deceptive calm. This is followed by a change in wind direction, so you need to move to the other side of your form of shelter. Move away from the coast if this is where you are situated, as a hurricane is often accompanied by tidal waves. If you are in a tent, take it down. Try to find shelter in a cave, on the lee side of an outcrop or dig a trench. Otherwise, find a solid building with a cellar and close all windows.

Tornadoes, or 'twisters', are small-volume cyclones. Most are relatively weak and register between F0–F3 on the Fujita scale, developing speeds of up to 320kph (200mph). Technically, an F6 rating – with speeds of around 610kph (380mph) – is possible, albeit highly unlikely. However, although severe twisters are rare, they do cause much more damage, as well as a higher percentage of tornado-related deaths.

The destructive capacity of tornadoes lies in their funnel tip – measuring only 20–50m (60–160ft) at the ground. The funnel fills with air, creating substantial pressure differences capable of lifting large objects such as cars into the spout. Do everything you can to stay away from them. Try to shelter in the most solid building you can find and close openings on the side of the twister while opening windows on the opposite side. If you are outdoors, lie flat in a ditch or holding onto the base of a tree.

VIOLENT WINDS WHIRLING AROUND A SMALL LOW PRESSURE AREA CREATE THE CHARACTERISTIC FUNNEL SHAPE OF A 'TWISTER'.

Index

Note: page numbers in **bold** refer to illustrated material.

Photographic credits

Cover Mountain Camera/John Cleare

SIL = Struik Image Library

2	The Picture Box/Steve Turner
4-5	The Picture Box/Uli Himsl
7	SuperStock
8a	Bill Hatcher
9	Gallo Images/Tony Stone
10	Paul Harris
11	Hedgehog House/Colin Monteath
12	SIL/Nick Aldridge
13	Auscape
14	Picture Box/Phil Schermeister
15	Auscape/Jan-Peter Lahall
16	Gallo/Tony Stone
17a	SIL
17b	SIL
17c	SIL
17d	SIL
17e	First Ascent
17f	SIL
17g	SIL
17h	SIL
17i	SIL
18	Petzl
19a	SIL
19b	SIL
19c	SIL/Ryno Reyneke
20	SIL/Jacques Marais
21	SIL/Jacques Marais
22a	SIL
22b	Hedgehog House/Paul Rogers
23	David Bunnell
24	The Picture Box/Hoa Qui
25a	SIL/Jacques Marais
25b	Auscape/D. Parer & E Parer-Cook
26a	KOS/Gilles Martin-Raget
26b	SIL
27	SIL/Jacques Marais
28	Spectrum Stock/R Mackinlay
29	Picture Box
30	Stockshot/Jess Stock
32c	Glenn Randall
33	Marie Lochman/Lochman transparencies
34	Hedgehog House/Barbara Brown
37b	Colin Monteath/Auscape
38	Raytheon Marine Company
41a	SIL
41b	Heather Angel
41c	Auscape/Wayne Lawler
42a	Picture Box/Hoa Qui
43	Hedgehog House/Peter Cleary
44	SIL/Jacques Marais
44b	Auscape
46	SIL/Jacques Marais
47a	Hedgehog House/Walter Fawlie
48	Picture Box
49	Onne van der Wal
50a	Colin Monteath/Auscape
50b	Hedgehog House/Dick Smith
51	Auscape/D. Parer & E Parer-Cook
53d	Gallo Images/Tony Stone
54	SIL/Jacques Marais
55a	Richard Sale
56	RSPCA Photo Library/ Peter Gasson
57a	Photo Access
57b	Andy Belcher
58a	SIL/Jacques Marais
58b	Photo Access
60	SIL/Ryno Reyneke
61a	SIL
61b	SIL
66	Gallo Images
67	SIL/Jacques Marais
69a	Dave Davies
69b	SIL/Jacques Marais
72	Gallo/Tony Stone
73	Stockshot/D Willis
74a	SIL/Jacques Marais
74b	Hedgehog House/Chris Rudge
78	SIL/Jacques Marais
78b	SIL/Jacques Marais
79	Anders Blomqvist (Seeing Eye)
80	SIL/Jacques Marais
81	SIL/Jacques Marais
82a	Gallo/Dugald Bremner
82b	SIL
83a	Gallo Images
83b	SIL
84	SIL/Jacques Marais
85	SIL/Nicholas Aldridge
86	Gallo/Tony Stone
87	Gallo/Tony Stone
88	Andy Belcher
89	Stock Shot/Jess Stock
90	Mountain Camera/John Cleare
91	FLPA/Terry Whittaker
92	Pictures Colour Library
93	Photo Access

Morse code

A	• –	T	–
B	– •••	U	•• –
C	– • – •	V	••• –
D	– ••	W	• – –
E	•	X	– •• –
F	•• – •	Y	– • – –
G	– – •	Z	– – ••
H	••••		
I	••		
J	• – – –	1	• – – – –
K	– • –	2	•• – – –
L	• – ••	3	••• – –
M	– –	4	•••• –
N	– •	5	•••••
O	– – –	6	– ••••
P	• – – •	7	– – •••
Q	– – • –	8	– – – ••
R	• – •	9	– – – – •
S	•••	0	– – – – –

Leave a gap of a few seconds between letters (depends on your speed) and a bigger one between words.

AAA = END OF SENTENCE   AR = END OF MESSAGE

IMI = Do not understand – repeat   SOS = ••• – – – •••

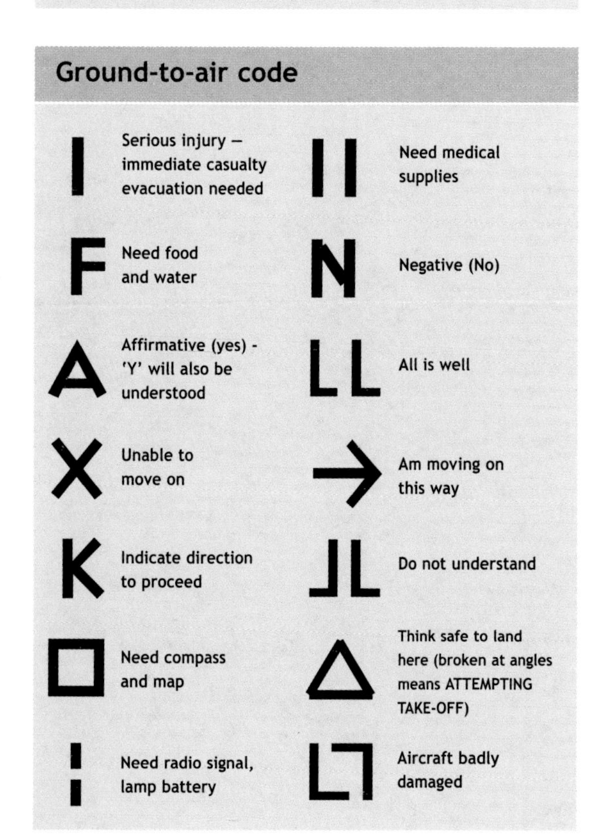

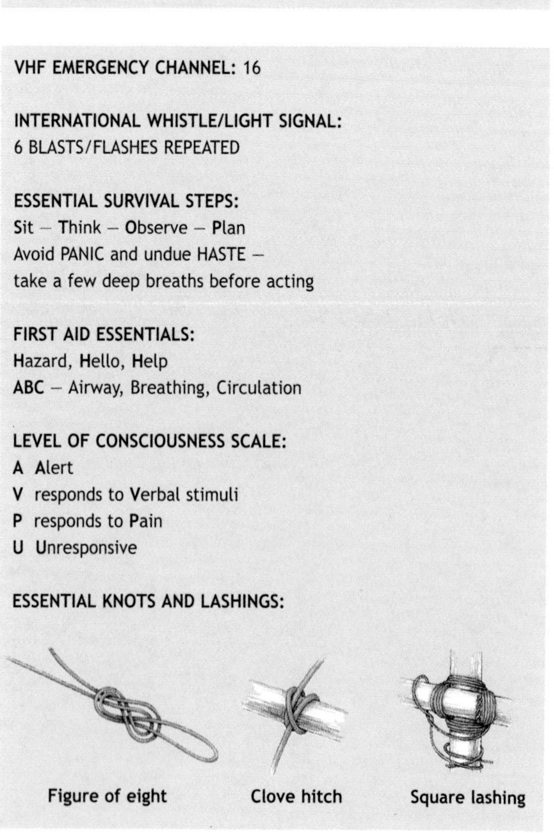

VHF EMERGENCY CHANNEL: 16

INTERNATIONAL WHISTLE/LIGHT SIGNAL:
6 BLASTS/FLASHES REPEATED

ESSENTIAL SURVIVAL STEPS:
Sit – Think – Observe – Plan
Avoid PANIC and undue HASTE – take a few deep breaths before acting

FIRST AID ESSENTIALS:
Hazard, Hello, Help
ABC – Airway, Breathing, Circulation

LEVEL OF CONSCIOUSNESS SCALE:
A Alert
V responds to Verbal stimuli
P responds to Pain
U Unresponsive

ESSENTIAL KNOTS AND LASHINGS:

Figure of eight   Clove hitch   Square lashing